Open Labs and Innovation Management

This book examines returns on experience and managerial practices to generate deeper collaboration, intensify co-creation, support start-ups and established companies to explore, develop, and accelerate their projects thanks to open labs (living labs, fab labs, coworking spaces, "third spaces", etc.). Open labs are the beatbox to create a rhythm in ecosystems and make all stakeholders move forward, faster, together. This book proposes a framework to understand how open labs, innovation hubs, and collaborative spaces contribute to ecosystems.

The book looks beyond the short-term effects of open labs and identifies four main dimensions: communities, physical spaces, events, and portfolios of services offered to private businesses, entrepreneurs, and start-ups, established companies, or public institutions. Drawing on extensive field research lasting over five years, with more than 40 cases and more than 200 interviews plus direct observation within different environments, this edited book investigates how managers run these labs, and how "users" or "clients" evolve when benefitting from their services. All chapters analyse how an actual management impacts the dynamics of communities, how it shapes the co-evolution between open labs and their ecosystems, and how the management of the physical space impacts the mission of the lab and its role in the ecosystem.

Open Labs and Innovation Research is written for scholars and researchers in the fields of innovation studies and management science. This book can also inform teaching, public policymaking, and professional practice.

Valérie Mérindol is Professor at the Paris School of Business. She teaches the management of creativity and innovation and also knowledge management.

David W. Versailles is Professor at the Paris School of Business, lecturing in strategic management, innovation studies, and business modelling. David is also founding partner and CEO in ISK Consulting SA (Luxembourg), and a visiting professor at Luxembourg School of Business.

Together, Valérie and David co-head the PSB New Practices for Innovation and Creativity (newPIC) chair, which specialises in the investigation of the micro-foundations of innovation and creativity.

RIOT!

Routledge Studies in Innovation, Organizations and Technology

Disruptive Platforms
Markets, Ecosystems and Monopolists
Edited by Tymoteusz Doligalski, Michał Goliński and Krzysztof Kozłowski

Inclusive Innovation
Robyn Klingler-Vidra, Alex Glennie and Courtney Savie Lawrence

Organizational Change, Innovation and Business Development
The Impact of Non-Technological Innovations
Edited by Magdalena Popowska and Julita E. Wasilczuk

Public Innovation and Digital Transformation
Edited by Hannele Väyrynen, Nina Helander and Harri Jalonen

Innovation and Leadership in the Public Sector
The Australian Experience
Mahmoud Moussa, Leonie Newnham, Adela McMurray and Nuttawuth Muenjohn

Advancing Big Data Analytics for Healthcare Service Delivery
Tiko Iyamu

Open Labs and Innovation Management
The Dynamics of Communities and Ecosystems
Edited by Valérie Mérindol and David W. Versailles

For more information about this series, please visit: www.routledge.com/Routledge-Studies-in-Innovation-Organizations-and-Technology/book-series/RIOT

Open Labs and Innovation Management

The Dynamics of Communities and Ecosystems

Edited by
Valérie Mérindol and David W. Versailles

Routledge
Taylor & Francis Group
LONDON AND NEW YORK

First published 2023
by Routledge
4 Park Square, Milton Park, Abingdon, Oxon OX14 4RN

and by Routledge
605 Third Avenue, New York, NY 10158

Routledge is an imprint of the Taylor & Francis Group, an informa business

British Library Cataloguing-in-Publication Data
A catalogue record for this book is available from the British Library

ISBN: 978-0-367-61278-8 (hbk)
ISBN: 978-0-367-64639-4 (pbk)
ISBN: 978-1-003-12558-7 (ebk)

DOI: 10.4324/9781003125587

Typeset in Bembo
by Deanta Global Publishing Services, Chennai, India

Contents

Figures

Tables

Contributors

Carl-Éric Aubin is a full professor in the Department of Mechanical Engineering at Polytechnique Montréal, a researcher at CHU Sainte-Justine and an associate professor in the Department of Surgery at the Faculty of Medicine of the Université de Montréal. Prof. Aubin holds the NSERC/Medtronic Industrial Research Chair in Spinal Biomechanics. He is a fellow of the Canadian Academy of Engineering and an Officer of the Order of Canada.

Carl-Éric Aubin is Executive and Scientific Director of the TransMedTech Institute which he founded. This Institute, supported by the Canada First Canada Research Excellence Fund, aims to support the development of innovative medical technologies and to train the next generation of professionals in this field.

Carl-Eric is an expert in spinal biomechanics and in the optimisation of orthopaedic treatments. He has published over 290 articles in scientific journals or edited books and over 690 abstracts in scientific journals or conference proceedings. He is the co-author of seven patents.

LinkedIn: www.linkedin.com/in/carlericaubin/

Nicolas Aubouin holds a PhD in management sciences from Mines ParisTech. His research focuses on the management of arts, of creativity, and on human resources management. He works more specifically on dynamic capabilities, collaboration, the organisation of "third places", institutional processes, and the evolution of artists' roles and competencies.

Nicolas is currently an associate professor in Management and Organisation Science at Paris School of Business where he was previously the head of the research department on Strategy and Management. He is an active contributor to the newPIC chair, and also a research associate in the CGS Research Centre at Mines ParisTech. Nicolas closely collaborates with public and private institutions for his research, primarily in relation to arts and culture.

LinkedIn: www.linkedin.com/in/aubouin-nicolas-15234b11/

Ignasi Capdevila is a full professor at Paris School of Business, and an associate researcher at Mosaic (the Creativity and Innovation Hub at HEC

Montréal) and at BETA (University of Strasbourg, France). He is an active contributor to the newPIC chair. Ignasi holds a PhD in Management from HEC Montréal and the HDR (French Habilitation) from the University of Strasbourg. He also holds three engineer diplomas, and an Executive MBA from ESADE Business School. He worked in the automotive industry for more than 12 years before entering academia, with managerial positions in R&D departments in several companies in European countries.

Ignasi's research interests include localised knowledge dynamics, knowledge communities, and creativity and innovation management in organisational and urban contexts. Ignasi is currently working on innovation dynamics in collaborative spaces (like coworking spaces, fab labs, makerspaces, hackerspaces, etc.) and on the knowledge dynamics and creative and innovation processes taking place in cities and creative industries.

LinkedIn: www.linkedin.com/in/ignasi/

Patrick Cohendet is a full professor at HEC Montréal in the International Business Department. His research interests include Theory of the Firm, Economics of Innovation, Economics of Knowledge, Economics of Creativity, and Knowledge Management.

Patrick is the author of 20 books and over 120 articles in edited journals, such as *Research Policy*, *Organization Science*, *Industrial and corporate Change*, *Journal of Economic Geography*, *Long Range Planning*, etc. He was the supervisor of more than 80 PhDs. He conducted a series of economic studies on the economics of innovation for different international organisations such as the European Commission, the Council of Europe, the European Space Agency, and the Canadian Space Agency.

Patrick is co-director of the research group Mosaic at HEC Montréal on the management of innovation and creativity, and co-editor of the academic journal *International Management*.

LinkedIn: www.linkedin.com/in/patrick-cohendet-8245421/

Geneviève Cyr is currently a doctoral candidate in Public Health (specialisation in health systems, organisations and policies) at the University of Montréal. She has worked for the last three years as a co-project coordinator of an interdisciplinary living lab in the imaging and orthopaedic research lab (Laboratoire de Recherche en Imagerie et Orthopédie) (LIO) affiliated with the Higher School of Technology (École de Technologie Supérieur) (ETS) and the Research Centre of Montréal University Hospital (Centre Hospitalier Universitaire de Montréal) (CR-CHUM). She is also a research assistant at the Mosaic research group on innovation and creativity management at HEC Montréal.

Geneviève's research focuses on open innovation in healthcare, more specifically living labs, sustainable innovation, and user engagement. She is interested in addressing complex problems within the healthcare system that require interdisciplinary collaboration to facilitate adoption of innovations, deliver greater value across the healthcare sector and improve the patient's experience.

LinkedIn: www.linkedin.com/in/gene-cyr/

Karl-Emanuel Dionne is an assistant professor at HEC Montréal in the Department of Entrepreneurship and Innovation.

Karl-Emanuel's work focuses on new approaches to organising innovation, cross-domain collaboration within and across organisational boundaries, open innovation events such as hackathons, bootcamps, and design thinking workshops. His research interests lie in how these new forms of organising help drive the development of new ecosystems. He currently studies these processes through qualitative methods in the emerging field of digital health, with start-ups, academic hospital organisations, and non-profit organisations.

LinkedIn: www.linkedin.com/in/karl-emanuel-dionne-a3999554/

Marie-Pierre Faure is Deputy Director at the TransMedTech Institute.

Marie-Pierre holds a PhD in Neuroscience from McGill University in Montréal and is the author of several scientific articles and books in the health field.

During her career, she has succeeded on several occasions in moving an innovative idea to the market with patented and registered products, both clinical and commercial. She is also a mentor to start-up entrepreneurs. Marie-Pierre has been building bridges between academic "knowledge" and "know-how" and innovative companies in the field of life sciences using the living lab approach (open innovation).

LinkedIn: www.linkedin.com/in/marie-pierre-faure-95ba121b/

Olivier Irrmann is Professor of Innovation Management and Co-Design at Junia, the engineering school at the Catholic University of Lille, France. He is the head of research for the strategic initiative on Innovation and Co-Design (ADICODE), a major project funded by the European Regional Development Fund and the Lille metropolitan area.

Olivier is developing research on collective intelligence, collaborative innovation, the management of creative hubs, and intercultural communication in business contexts.

Olivier holds a PhD in Management and Economics (KTT) from the Helsinki School of Economics (now Aalto University School of Business). He was working previously at Aalto University in Finland and HEC Montréal in Canada. Olivier is a research associate at the Mosaic Creativity and Innovation Hub based at HEC Montréal and coordinator of the Research Group on Collaborative Spaces (RGCS) for the Lille chapter.

LinkedIn: www.linkedin.com/in/olivierirrmann/

Alexandra Le Chaffotec is currently an associate professor at Paris School of Business. She holds a PhD in economics from University of Paris Sorbonne. She teaches economics and organisation studies. She is an active contributor to the newPIC chair at Paris School of Business. Before joining Paris School of Business, Alexandra first worked as a consultant with Ernst & Young; she was in charge of litigation, the evaluation of industrial property, and market analysis.

Alexandra focuses her research on on governance issues, the management of innovation, and the management of knowledge. Her research mainly applies to the field of healthcare-related organisations.

LinkedIn: www.linkedin.com/in/alexandra-le-chaffotec-201573218

Margaux Manent is currently a doctoral student in management, strategy, and organisations at HEC Montréal. She takes part in the "Innovators in Residence" research-action programme at CHU Sainte-Justine, which aims to support the development and implementation of a health technology innovation coming from an academic new venture. Working with Mosaic for two years, she is interested in hybrid forms of collaboration between institutions and the role of intermediary organisations in healthcare ecosystems.

LinkedIn: www.linkedin.com/in/mmanentd/

Valérie Mérindol is currently a full professor in Innovation and Creativity Management at Paris School of Business (PSB) where she co-heads the "new Practices for Innovation and Creativity" (newPIC) chair.

Valérie holds a PhD in political science from the University of Strasbourg and the Habilitation to supervise research in management (University of Strasbourg). She is also a founding partner in ISK Consulting SA (Luxembourg) and contributes to missions in relation to her research interests.

Valérie first worked for Crédit Agricole Indosuez as an analyst in charge of the aerospace domain. She joined then the Observatoire Économique de la Défense in the directorate for financial affairs of the French Ministry of Defence (MoD) as an expert on industrial and innovation policies. Between 2003 and 2008, she served as a senior researcher and as the deputy director in the French Air Force research centre. In this position, she headed a research team investigating organisational and managerial issues in relation to defence and aviation programmes. Valérie's last assignment before joining PSB was with the French governmental office Observatoire des Sciences et des Techniques, where she oversaw the evaluation of science policy, and of the governance of innovation. She was also an expert for the French HCERES. Valérie also contributes on a regular basis to expert groups in the French MoD and at inter-ministerial level in France.

Valérie's research interests include the management of innovation and creativity, innovation and science policymaking, governance issues, knowledge management, and the management of creativity.

LinkedIn: www.linkedin.com/in/valeriemerindol/

Octave Niamié holds a PhD in administration jointly offered by l'École des Sciences de la Gestion de l'Université du Quebec à Montréal (ESG-UQAM),

Concordia, HEC Montréal and McGill. He also holds a Master of Business Administration (Executive MBA) jointly offered by ESG-UQAM and PSL-Université Paris-Dauphine.

His research interests focus on venture capital, the entrepreneurial process, legitimacy, liminality, and collaborative spaces in entrepreneurship (incubator, accelerators, coworking). During his PhD programme, Octave won best communication awards in the student category at the 13th International Francophone Congress in Entrepreneurship and SMEs (CIFEPME) in Trois-Rivières in 2016, and at the eighth congress of the International Conference of the Association of Global Management Studies (ICAGM), in Montréal in 2018.

Octave is an assistant professor at Polytechnique Montréal and an active member of the Montréal entrepreneurship ecosystem, acting as board member or mentor with several entrepreneurship support organisations. He has co-founded several start-ups.

LinkedIn: www.linkedin.com/in/octaveniamie

Laurent Simon is a full professor of Creativity and Innovation Management at the Department of Entrepreneurship and Innovation, HEC Montréal Business School. He is co-director of Mosaic, a research hub on creativity and innovation. He received his PhD in administration from HEC Montréal.

His research interests include the analysis of the determinants of creativity in the management of innovation, with a focus on knowledge, management of ideas, and routine work. He conducted a series of studies on the management of ecosystems of innovation and of creative industries, in particular in the domain of videogame activities.

LinkedIn: www.linkedin.com/in/laurent-simon-905855/

Luc Sirois is an experienced executive, entrepreneur, and investor in digital technology and healthcare innovation. With a degree in electrical engineering from McGill University and an MBA from Harvard Business School, he was the Executive Director of Prompt, an industrial research consortium fostering and funding R&D partnerships in information technology, telecommunication, artificial intelligence, and cybersecurity. He is currently the Chief Innovation Officer at Québec Ministry of Economics and Innovation, and Executive Director, Québec Innovation Council.

His research interests and professional practice focus on new approaches to building innovation ecosystems, R&D collaborations, as well best practices in terms of co-creation events and innovation hubs design and management, all with a view to maximising their impact on economic development. Luc is known for his creative approach to innovation and for his leadership in digital technology. He is a passionate driver of innovation, creativity, and positive change wherever he gets involved. He co-founded Hacking Health, a digital health innovation movement now present in over 20 countries, along with a number of digital health innovation acceleration initiatives like the Hacking Health Accelerator.

Luc began his career in telecommunications at Bell Canada, and then joined McKinsey & Company as a strategic advisor to large corporations and institutions. He then co-founded the radiation oncology company Resonant Medical, now Elekta Canada, leading this successful enterprise through all stages from McGill University Health Centre spin-off start-up to world-class manufacturer and technology provider. He has also served as general manager or vice-president at Telesystem, TELUS Health, and Nightingale Informatix.

Luc is actively involved with multiple start-ups and non-profits in the areas of youth, healthcare, science, and education.

LinkedIn: www.linkedin.com/in/lucsirois/

Nathalie Tremblay currently holds the office of president of Gestion Névé Inc, is a researcher at HEC Montréal, Mosaic – Creativity and Innovation Hub in Canada and a doctoral candidate at the University of Strasbourg, Bureau of Theoretical and Applied Economics (BETA) in France. Nathalie holds an MBA in Management of International Affairs. She focuses her research on innovation studies in life science and medical technologies, and their integration in the healthcare system. Furthermore, she is also an expert consultant in the strategic management and impact of innovation.

Nathalie holds a strategic governance position for the Ministry of Economy and Innovation of Quebec, at the Quebec Research Fund – Health (FRQS), as a member of the board of directors and chair of the audit committee for a second term. The FRQS is responsible for the research strategies for health and life science for the Quebec government. Nathalie sits as a member of the board of directors of the Montréal Symphony Orchestra Foundation, managing the capitalisation funds of the orchestra. She sat as a member of the board of directors of the McPeak-Sirois Breast Cancer Clinical Research Group, a consortium of clinical trials including Quebec Research Institute and majors hospitals.

As a socially engaged entrepreneur for over 25 years, in the last decade, Nathalie was also CEO of the Quebec Breast Cancer Foundation and financing and funding director for the Montréal Symphony Orchestra.

LinkedIn: www.linkedIn.com/in/nathalie-tremblay-mba-5ba739124/

David W. Versailles is currently a full professor of Strategic Management at Paris School of Business where he co-heads the "new Practices for Innovation and Creativity" (newPIC) chair with Valérie Mérindol, and a visiting full professor at Luxembourg School of Business. David holds a Doctorate and the Habilitation to supervise research in economics (at University Aix Marseille 3 and University of Nice respectively). David has been also elected the SIG chair of the Strategic Interest Group (SIG) on Innovation for the European Academy of Management (EURAM) for the 2022–2025 term.

David was the chief scientist for the Directorate for Financial Affairs in the French Ministry of Defence and for the Conseil Économique de la Défense. David was then the founding director of the multidisciplinary research centre of the French Air Force. In this period, David also served as the first dean for research at the French Air Force Academy. After leaving the Ministry of Defence, he became a partner in consulting companies based in Brussels and Luxembourg where he developed a portfolio of activities in business strategy and business modelling. In November 2017, he created the consulting firm ISK Consulting SA in Luxembourg. David is currently the managing director of this company, and active at the same time as faculty and as a consultant.

David elaborates on the combination of economics, strategic management (including business modelling), industrial organisation, organisation theory, knowledge theory, and management of technology. He is active in these domains as principal investigator and project manager for research, teaching, and consulting. His research also covers the methodology of the social sciences, and the history of economic thought for the period 1870–1945, with a special focus on the Austrian School of Economics. David has worked extensively on civilian and military aeronautics, on safety and security issues, and on network centric organisations.

LinkedIn: www.linkedin.com/in/versailles/

Foreword

Elise Tissier
Director, Bpifrance Le Lab

Translation by David W. Versailles

The support to innovative businesses represents the kernel of Bpifrance's mission since its creation. The French government has tasked the Public Bank for Investments (Bpifrance) with the management of a vast programme (the "Deeptech plan") to support disruptive innovation and the creation of start-ups. However, innovation cannot be limited to the transformation of basic research born in academic laboratories. It should be considered from a broader perspective. This is the reason why Bpifrance has taken an interest in new organisational designs hosting all forms of innovation, technological or not. Such designs build a network of resources available all over France, and beyond. Are companies sufficiently aware of services offered by innovation platforms and open laboratories? Are managerial processes in place in companies adapted to interactions with these organisations? Are open labs suited to offer constant support to the development of innovation in companies? These questions were at the origin of interactions between Bpifrance and the newPIC chair at Paris School of Business; they motivated the research projects commissioned to this team of researchers.

Open labs as an "object" for entrepreneurs: what are we dealing with?

Local public policymakers have a vested interest in nurturing their territories, creating wealth, valorising, and retaining local businesses. They converge with innovative entrepreneurs on the need to work with innovation platforms and open innovation laboratories. These new organisational designs take multiple forms, but they are always intended to serve entrepreneurial ventures and innovative projects. They bring new solutions to support innovation, encourage new types of interaction, and promote new ways of working. They represent new links in local ecosystem networks. Bpifrance expected to improve its understanding of these new tendencies, and their roles in innovation ecosystems.

Seventy-four percent of managers explain that they intend to adapt their managerial practices over the coming years (Bpifrance Le Lab, 2020). Innovation platforms and open labs contribute to this evolution, and address some of the needs identified in companies. However, they do not only represent tools supporting the evolution of managerial practices and represent much more than convivial and disruptive open spaces. They also contribute to a deep transformation of social interactions in companies.

Understanding how to interact with innovation platforms and open labs is no easy task. What are the rationales for their installation? For their management? For their facilitation of innovation projects? For their support to incubation and acceleration processes? Bpifrance's mission is to support entrepreneurs in small and medium-size companies with the identification of available resources for their innovative ventures. A better understanding of the contributions by innovation platforms and open labs therefore makes it possible for Bpifrance to improve its own services, and the execution of its own mission. It is also important for Bpifrance to anticipate new tendencies and follow evolutions over time. Thanks to contributions by academic researchers, Bpifrance Le Lab made it possible to improve its understanding of developments occurring over recent decades and reinforce its raison d'être: "serve the future".

Essential contributions by academic research

Interactions with Paris School of Business's newPIC chair were motivated by the need for an analysis going much beyond a simple list describing existing innovation platforms and open labs. Bpifrance was looking for a taxonomy of such organisational designs that could be used to follow these intermediaries of innovation over time. Research projects developed by the team led by Valérie Mérindol and David W. Versailles at the newPIC chair of Paris School of Business were commissioned in partnership with Innovation Factory, a Paris-based innovation platform (also part of Galileo Education group) interested in understanding this then emerging phenomenon.

The first investigation (Mérindol and Versailles, 2017) made an essential and innovative contribution with the first taxonomy of open labs, innovation platforms, and "third places". Field research focused on the Paris region. The taxonomy de-homogenised these notions and identified several forms of innovation intermediaries suiting needs and expectations introduced by several categories of entrepreneurs, and of innovation projects. The diversity of interactions in ecosystems and of constitutive elements leading to the installation of innovation platforms was present in the taxonomic approach. Highly detailed information was collected during field research (interviews, observation), which made it possible to account for the large variety of interdependencies. This rich and dense material revealed all interdependencies and did not treat open labs as isolated agents. This systemic approach was the very reason for Bpifrance to commission the newPIC chair with this research project.

After this first round of activities, it soon appeared relevant to expand this research with a second report covering other regional ecosystems in continental France (Mérindol et al., 2018). A third report (Mérindol and Versailles, 2019) then compared the different evolutions in France and in different Asian countries. The Paris region may be the sole French "innovation leader" region on the European regional innovation scoreboard (2016), but it is neither the only French region to host innovation platforms, open labs, and "third places", nor the sole region to host innovation-related initiatives (in the different categories of innovation). Open labs exist in all French regions. New innovation platforms emerge every day. Interactions between local policymakers and companies foster their multiplication, and their diversity. It is then highly relevant for Bpifrance Le Lab to follow these tendencies and take advantage of this knowledge to feed other research projects and field research activities.

Lessons learned

Open labs serve the installation of new organisational designs in companies

The diversity of fab labs, incubators, makerspaces, coworking spaces, and other organisational designs in relation to innovation was initially difficult to appraise. With their different objectives, services, and ways of working, their potential contributions to entrepreneurial ventures were so diverse that it was difficult to identify their respective added values. The contributions of academic research made it possible to better understand how innovation platforms and open labs support the ever-changing dynamics of entrepreneurial ventures and ecosystems, both established companies and SMEs. They offer complementary and diversified competences and distinctive experiences. Open labs serve communities of entrepreneurs where horizontal relations abound. New modalities prevail for the interactions, beyond statuses (employees or not), hierarchical links, or roles in organigrams. Strong ties emerge and foster the dissemination of sharing and reciprocity values. Open labs also represent a source of creativity. They project a positive value of entrepreneurship in ecosystems that concretises with the creation of new businesses. This renewed environment also contributes to the shedding of a different light on entrepreneurship, where failure is acknowledged as a standard step in any entrepreneurial journey where teams and individuals learn. Failure is not considered as an infamous definitive dead-end anymore; it represents the logical by-product of any entrepreneurial risk and the logical corollary of learning-by-making.

Open labs commit to several missions

The first main contribution provided by the taxonomy of open labs and innovation platforms focuses on their services, with the important dichotomy between "thinkers" and "makers". Open labs adhering to the "maker" approach develop a portfolio of activities around prototyping. "Thinker" open labs deliver intellectual services in relation to innovation and creativity. Beyond these two categories, open labs share several features: a physical space, communities of individuals and entrepreneurs, open and collaborative approaches, and an openness to all stakeholders in their respective ecosystems. Open labs define themselves by their portfolio of services and missions, and by the collective dynamics they foster, develop, or contribute to. It is especially interesting to understand how open labs contribute to start-up creation and business ventures. Some of them focus on a specific sector, others do not. Some organise to support independent workers and craftsmen with the creation of competence networks to mutualise resources and produce joint responses to calls for proposals. The main important societal topics of our time (social and solidary economics, economic and environmental sustainability, "civic techs") frame activities in lots of open labs. Other open labs elaborate on geeks passionate about technology who commit to developing their projects during their free time. The internal organisation of open labs always demonstrates their originality, with facilitation staff, communities, and mentors in charge of incubation. Whatever their size, small open labs or mega-innovation platforms, all these aspects finally materialise with the emblematic spaces embodying their brand. In all regions, open labs also develop with transport and communication infrastructures, with stakeholders in the ecosystems, and with the other activities present in cities and urban environments.

Open labs provide new opportunities for developing firms' open innovation strategy

Investigations developed by Bpifrance show that the foundations of innovation lie in the confrontation of ideas between services, units, etc. Innovation is always intended as a process leading to value creation. Such confrontation represents a condition for creativity. It is also a reference in the process leading to a go/no-go decision in the innovation process. All activities in place in companies with collaborative processes, decentralised responsibilities, and horizontal participation do not automatically lead to innovation. However, cooperation and the sharing of ideas and returns on experience always represent guarantees for innovation and performance. Fruitful collaboration paves the way for organisations to economise on time and cash (Bpifrance Le Lab, 2020). Research also shows that community-based communication, for instance with internal social networks available in companies, supports interactions about ideas and knowledge, and contributes to creative stimulation.

Following the same rationales, open labs and innovation platforms represent additional and complementary resources worthy of interest for any company. Open labs offer an opportunity to externalise parts of a company's human resources beyond its traditional boundaries, on a permanent or temporary basis. It is relevant to envision open labs as extensions of companies connecting with new networks of stakeholders and contributors of innovation and creativity projects. Research developed by the newPIC chair clearly identifies that "communities represent the key distinctive assets in open labs and innovation platforms", "a source of diversity of experience, competencies, and perspectives". Open labs elaborate on these communities to provide new intermediation services in the general framework of open innovation and facilitate adaptation to change in always more turbulent business environments.

Whatever their size, firms are today both clients and partners of open labs. As clients, firms take advantage of new opportunities based on new intermediation capabilities proposed by open labs and, also, purchase their services in the domain of innovation and creativity. As partners, firms commit to long term interactions that are not only based on commercial rationales. This dual relation shows how much it is relevant to investigate further the new framework by open labs. All categories of firms, in different sectors, with different levels of technological intensity, with different strategic intents and different objectives, will find opportunities to benefit from interactions with open labs. It applies to small companies and entrepreneurial ventures that will find options in open labs to reach out to a large population of stakeholders and grow faster, and in more solid ways, in their ecosystems. It also applies to large established companies that will be supported by open labs to find flexibility and agility lost because of their size. Research on open labs is important because these organisations represent key actors in business ecosystems.

Open labs support firms to cope with increasing levels of complexity

Among many other variables, globalisation and digitalisation are major elements reshaping the world, illustrating its complexity. Managers and strategic decisionmakers need to create the conditions for the firm's agility to adapt. This constraint applies to all sorts of entrepreneurial ventures, and to all categories of companies, but firms are obviously less agile when their size and complexity increases. In these cases, creativity and innovation must find ways and means to transform into actual evolutions of business models, strategies, or portfolios of products/services. Managers cannot rely on linear modalities anymore: they need to find options to adapt to this complexity, and cope with it to sustain competitive advantages in their organisations and empower innovation processes. Managers shall transform with exceptional competencies to handle the different levers available in firms that can be presented as a sort of complex adaptative system (Heraud et al., 2019). Whatever the theoretical references at hand, the management of creativity and innovation does not fit into models of the firm based on strategic planning and linear reasoning.

New learning processes, new forms of interaction, of cooperation, of sharing, make it possible to address the challenges of complexity. In taking advantage of the dynamics of communities, research about open labs and innovation platforms has identified how these new intermediaries install specific mechanisms fostering innovation and creativity that propose actual solutions to complex problems thanks to dense attention to end users, observation, and communication. Research is still necessary to investigate further how to deliver and manage open labs and innovation platforms, but it is already possible to conclude that these aspects represent key distinctive contributions made by open labs to ecosystems.

Open labs make direct contributions to local economic policies

Last, but not least, services delivered by open labs and innovation platforms serve public policies. This explains why local, national, and European public policymakers not only devote increasing attention to them, but also support their development. Local public actors, municipalities, or regional governments do not only bring budgets, but also develop specific innovation programmes based on open labs. This "political" contribution represents an important evolution. Local policymakers and "political" entrepreneurs partner to establish specific open labs or use services intended to support the development of start-ups and SMEs, societal transformation, and the development of local business ecosystems. As soon the open lab gains influence on a territory, it also becomes a part of its public brand and contributes to its global visibility and attractiveness.

France is also characterised by the emergence of mega open labs, or mega-innovation platforms, that elaborate on specific financial resources and directly operate on a very large scale attracting international visibility. Some of these ventures become platforms of platforms and deserve specific attention to better understand the services delivered to communities of entrepreneurs and firms. Research has already identified the tension between size (or scale) and the preservation of strong ties inside communities. Further research is probably necessary to investigate these aspects further and qualify how to preserve the philosophy of "sharing" and reciprocity that constitutes the originality of open labs.

Perspectives for the future

Thanks to rigorous academic research and well-documented field research, Bpifrance Le Lab is now able to support SMEs and explain the open labs' added value for the development of entrepreneurial trajectories, and for activities in relation to exploration and innovation. Bpifrance is committed to supporting the development of strong ties between open labs, SMEs, and medium-sized industrial companies. It is the companies' vested interest to embark on this evolution.

However, managerial modes must adapt. To maximise their attractiveness, open labs and innovation platforms must preserve flexible managerial modes and minimal hierarchies for the facilitation of projects and avoid the traps of unnecessary constraining structures. New management modes also emerge within companies with improved facilitation mechanisms and greater consideration for nonlinear processes of innovation. Open labs and innovation platforms have understood this transition well, but mega open labs will devote great attention to this issue in the future. This point is important because the world of open labs will most probably reveal restructuring in the future, with more sustainable business models and stabilisation around high value added services.

Bpifrance Le Lab aims to provide SMEs and mid-size companies the keys for their expansion, growth, and development to contribute to national influence. This research on open labs and innovation platforms contributes to our mission.

References

Bpifrance Le Lab. (2020). *Sens et liberté. Revenir aux fondamentaux du management.* Paris: Bpifrance. https://lelab.bpifrance.fr/Etudes/sens-et-liberte-revenir-aux-fondamentaux-du-management.

Bpifrance, Deeptech Plan. http://www.bpifrance.fr.

European Commission, Directorate-General for Internal Market, Industry, Entrepreneurship and SMEs, *Regional Innovation Scoreboard 2016, European Commission, 2017*, https://data.europa.eu/doi/10.2873/84730.

Héraud, J.-A., Kerr, F., and Burger-Helmchen, Th. (2019). *Le management creatif des systèmes complexes* (Vol. 20). London: ISTE Editions.

Mérindol, V., and Versailles, D. W. (2017). *Créer et innover aujourd'hui en Île-de-France : le rôle des plateformes d'innovation.* Projet de recherche financé par Bpifrance Le Hub, Innovation Factory et Paris&Co (avec des contributions d'Ignasi Capdevila, Alexandra Le Chaffotec, Nicolas Aubouin, Marion Desnost et Marianne Cohen pour la recherche de terrain). Paris: PSB.

Mérindol, V., and Versailles, D. W. (2019). *Créer et innover aujourd'hui en France et en Asie : le rôle des plateformes d'innovation et des open labs d'entreprise.* Projet de recherche financé par Bpifrance Le Lab et Innovation Factory. Paris: PSB.

Mérindol, V., Versailles, D. W., Le Chaffotec, A., Aubouin, N., and Capdevila, I. (2018). *Créer et innover aujourd'hui en France : le rôle des plateformes d'innovation dans les écosystèmes régionaux.* Projet de recherche financé par Bpifrance Le Lab et Innovation Factory (avec des contributions d'Oceane Duyck et Salim Moulmaaz pour la recherche de terrain). Paris: PSB.

Acknowledgements

We express our gratitude to all open labs that have hosted our field research activities over the years, and which accepted to organise interviews, visits, observation activities, focus groups, and seminars with us. They represent the foundation for this book.

Our attention to open labs emerged with an expert group that we co-organised with the FutuRIS platform at ANRT in 2015. The editors are indebted to Bpifrance Le Lab and Innovation Factory for the partnership with the newPIC chair at Paris School of Business that made it possible to expand this investigation with several research projects, and for the research grants supporting a significant part of the research presented here. On a personal level, we would like to express our gratitude to Anne Lalou, Elise Tissier, Marie-Laure Henry, and Frédérique Savel for their support in developing a parallel journey into the investigation of concepts and the appraisal of managerial challenges present in the open lab phenomenon.

We also express our gratitude to Michel Ida for sharing with us his long experience of open labs. Michel developed open labs before anybody else in France and demonstrated a visionary perspective on this movement. The Afterword in this book shows, once again, that Michel is one step ahead in the analysis of the management of innovation.

We thank all the authors for their contributions, ideas, and patience in the elaboration of this book. We are specially indebted to Patrick Cohendet, for his encouragement and companionship over all these years. Only Patrick's support made it possible to expand the ambition of this book beyond France.

During the difficult months of the pandemic, we had the opportunity to organise a digital workshop to discuss our respective contributions. Let's hope that the months to come will give us the opportunity to meet during physical events and enjoy many more discussions on this great topic.

We hope this book encourages more research on open labs and, in turn, causes the science of innovation intermediaries to evolve and mature.

Introduction

Valérie Mérindol and David W. Versailles

This book shows the importance of open labs as catalysts of innovation, or as innovation intermediaries, to handle mutations and challenges in knowledge-based economies.

Working spaces, living labs, fab labs, incubators, makerspaces, and hackerspaces are considered as new phenomena best suited to improve the management of creativity and of innovation (Merkel, 2017; Bouncken and Reuschl, 2018). These new workplaces and spatial constructs (Burkner and Lange, 2020), developed over the 2010 decade. Howell and Bingham (2019) identify around 14,000 active coworking spaces in the USA in 2017. In France, the number of coworking spaces has multiplied by a factor of ten over the last ten years. Fablab Studio assesses 250 active fab labs and 2,500 active makerspaces in 2020 in France. In China and in India, the same phenomenon does exist but with a twin focus on incubators and coworking spaces.

In this book, we call these new organisations "open labs" to zoom out from the different specificities introduced by labelling and certification networks (fab labs, living labs) and other societal claims (coworking spaces). Despite their diversity and various labels, they are all built on three pillars: a community of contributors, a physical space to host social interactions and knowledge processes, and a portfolio of services (such as an incubation programme, flex office, or consulting activities). Even though they share many properties and managerial challenges with open labs, this book deliberately leaves out innovation platforms affiliated with companies, because they respond to different rationales (business models, managerial practices, relations with intrapreneurship, reappropriation of innovation outcomes inside business units), and usually make an impact on business ecosystems through the adaptation of the business models and business portfolios of established firms.

The purpose of this book is to appraise the originality of open labs and the dynamics of their evolution. It offers the opportunity to investigate their multifaceted contributions to the management of innovation and creativity. The chapters explain that open labs require proactive managerial practices, with events, and adapted physical and organisational designs. To accommodate the demands and expectations introduced by stakeholders and users, mindsets must adapt. Open labs are the beatbox to create a rhythm in ecosystems and

DOI: 10.4324/9781003125587-1

make all stakeholders move forward, faster, together. To travel this journey, engage key stakeholders, and make an impact, new attitudes and skills are required to empower collective competences and support communities.

The originality of this book relates to the investigation of managerial issues, usually a black box or a set of implicit, often irrelevant, transpositions from other contexts. In all chapters, the main conclusion is simple: setting up a physical space or installing a portfolio of "open lab-related" activities without adapted management makes no impact. Actual management is a mandatory condition for success. The chapters analyse these managerial specificities and key success factors necessary to make an impact on ecosystems. The book explains how managers orchestrate all activities to deliver services to "users", and to create value through the empowerment of communities.

The book originated in a discussion between the editors when they were sitting under the sun of the Praça de Comercio in June 2019, between sessions of the European Academy of Management annual conference held in Lisbon, Portugal. The discussion soon reached out to our colleagues at the Mosaic research centre in Montréal. We all realised that our dense field research activities were contributing to propose an original light on this important phenomenon with a large sample of original cases that were accessible to English-speaking audiences with difficulty.

On the French side, activities started in late 2014 with an expert group organised and facilitated by the editors of this book for the newPIC chair at Paris School of Business, in partnership with the FutuRIS platform of the French Association Nationale de la Recherche et de la Technologie (ANRT). The publication of a white paper ensued, that was among the first publications on the topic in France (Mérindol et al., 2016). This document drew the attention of different institutional audiences in France, and motivated Innovation Factory (a Paris-based open lab) and Bpifrance Le Lab (the research lab of the French governmental institution supporting innovation, Bpifrance) to commission several research projects to the newPIC chair, all of them being directed and supervised by the editors. All projects compared open labs with innovation platforms hosted by (large) companies. The first research project investigated the open lab phenomenon in the Paris region (Mérindol and Versailles, 2017). The second one expanded the analysis to continental France's main regional hubs (Mérindol et al., 2018). The third investigation introduced updates about the phenomenon in France, but mainly compared the recent evolutions in France and in different countries in Asia (Mérindol and Versailles, 2019). For this last project, some aspects of field research activities were performed by *Innovation is Everywhere*, a Singapore-based consulting firm commissioned *ad hoc* for the project. In parallel, Ignasi Capdevila was focusing on specific aspects of coworking spaces, a specific category of open labs, in supervising and contributing to several research projects funded by the regional government of Catalonia in Spain (Cowocat Rural), or by the European Union (Coral project on collaborative spaces).

All over these years, the Mosaic team at HEC Montréal, headed by Patrick Cohendet and Laurent Simon, was organising several projects about local open lab initiatives, and promoting the originality of this phenomenon during the different editions of the Summer School on Management of Creativity organised in Montréal (Canada), Barcelona (Spain), or Strasbourg (France). Several authors or co-authors of this book are (were) PhD students in the Mosaic research team. They have investigated or are investigating open labs for their PhD projects.

As a follow-up to this dense field research endeavour operated over eight years, the book offers an outline of more than 40 open labs located in France, Canada, Spain, China, and other countries in Asia. The ten chapters in this book are all original contributions. Most case studies analysed in these chapters are published here in English for the very first time.

Open labs as a key player of open innovation

The phenomenon of open labs, innovation hubs (or platforms), and collaborative spaces is anchored in the general dynamics of open innovation, but it has now pervaded lots of different areas of business and economic life. Open innovation is considered as a new paradigm of innovation (Chesbrough, 2003; Bogers et al., 2018): because of the increasing complexity of the knowledge base and the turbulence of business and technological environments, public and private actors cannot innovate alone anymore. Large organisations cannot rely solely on internal resources to innovate. Smaller companies and start-ups must embed themselves in networks that become even more critical than before to find the resources appropriate to develop innovation projects or new ventures. In the context of open innovation, the combination of internal and external resources is necessary to explore, identify new opportunities, experiment with new solutions, and bring them to the market.

Two categories of consequences of this new paradigm ensue. First, open innovation implies collective strategies between public and private actors to develop new projects. Open innovation requires the development of ecosystems strategy (Jacobides et al., 2018) and to nurture various communities of innovation, interests, and practices (West and Lakhani, 2008; Amin and Cohendet, 2004; Roberts, 2017). Second, open innovation requires the connection of profiles with various backgrounds and different types of competencies. This implies the installation and development of the conditions of relational trust and common/mutual understanding to nurture new multisided collaboration (Ollila and Elmquist, 2011). When appraised from a knowledge management perspective, the exploration and exploitation of new solutions in an open innovation context requires the articulation of heterogeneous knowledge.

"Open labs" directly contribute to the management of creativity and innovation in the context of open innovation by designing new ecosystems and communities (Mérindol and Versailles, 2017). They make it possible to break silos by connecting people from various institutions and environments. Open labs progressively become the focal point in an economy of serendipitous encounters (Jakonen et al., 2017) and a locus of creativity in the context of globalisation (Bathelt and Cohendet, 2014; Capdevila, 2015; Suire, 2019). They also offer the opportunity to develop new managerial practices to support collaboration between public and private actors and to experiment new creative methods (Sarpong et al., 2017).

This book illustrates, with actual cases, recent tendencies and development trajectories. It investigates, in particular, their contributions to innovation ecosystems in sectors such as healthcare, smart cities, or tertiary education. The book also offers perspectives to explain how open labs progressively become focal actors in their respective ecosystems. In addition, it shows how open labs present a unique opportunity to install new connections with stakeholders who are not traditional parts of the innovation processes or entrepreneurial journeys, such as artists. Furthermore, the book also shows how open labs diffuse their organisational model from urban areas to rural environments.

Conceptual approaches of open labs as original intermediaries

From a conceptual perspective, open labs belong to the category of organisational intermediaries inside ecosystems. In line with the analysis of networks developed by Hargadon and Sutton (1997), Howells (2006) considers that intermediaries do not only act "linkers" but also as "brokers". It means that open labs actively participate in the creation of value in ecosystems, and progressively transform into a knowledge repository to develop new solutions. Agogue et al. (2013) suggest two complementary functions to investigate the role of these intermediaries: they act as "brokers of networks" and as "brokers of content" by offering various services to develop collective strategies and collaboration.

Several forms of organisational intermediaries exist (Howells, 2006; Agogue et al., 2013): firms, not-for-profit organisations, informal groups of individuals. In this book, we consider open labs as an original type of organisational intermediaries based on the dynamics of communities. They gather people coming from various economic spheres and contribute to progressively create a sense of community among people who do not know each other (Garret et al., 2017; Bouncken and Alsam, 2019): entrepreneurs, artists, designers, employees coming from large companies, and scientists. This book investigates open labs as specific instances of organisational intermediaries based on the dynamics of communities, and the installation of physical spaces for interactions. Open labs contribute to build new cognitive architectures by animating events, coaching for innovation and creativity methods, and organising collaboration. This book shows that open labs act as catalysts in new ecosystems of innovation and contribute to the rejuvenation of industrial wastelands in urban areas and peripheral regions.

Open innovation requires the installation of new models of governance at the local level to install new dynamics of knowledge production based on collective strategies. Open labs contribute to installing these new modalities of interaction and coordination modes among the main actors of ecosystems.

Because communities are located at the kernel of their development, open labs can be considered as promoters of three forms of governance. In this perspective, the open lab phenomenon locates at the intersection between three conceptual frameworks.

First, open labs represent a "*middleground*" as defined by Cohendet et al. (2014). It means that they characterise a link, an intermediary framework based on communities and on dynamics that are not only led by prices and commercial mechanisms, between an informal underground culture, and formal organisations (the "upperground"). As instances of the "middleground", open labs contribute to make visible (or available) ideas and solutions already present in the underground, and to transform them into innovation for the "upperground" (Bathelt and Cohendet, 2014; Schmidt and Brinks, 2017; Brown, 2017). This book shows that the concept of "middleground" is useful to understand how open labs contribute to knowledge intensive sectors such as healthcare ecosystems.

Second, open labs represent the trigger of the Triple and Quadruple Helix models of innovation (Etzkowitz and Leydesdorff, 2000; Carayannis and Campbell, 2009). The Triple Helix describes virtuous interactions and collaboration between policymakers, firms, and universities (Etzkowitz and Zhou, 2017). The Quadruple Helix extends the collaboration between these three institutional spheres to civil society, most notably citizens, artists, and media (Carayannis and Rakhmatullin, 2014). Open labs contribute to the emergence of a climate of trust suited to install the dynamics of the Triple or Quadruple Helix models of innovation at territorial level. They represent new spaces of interactions to create collaborative projects and to build a common picture of priorities that is shared by public and private actors (Ranga and Etzkowitz, 2013; Heraud, 2017). Open labs also contribute to experiment in new ways to involve citizens, artists, and "normal" users in creativity or innovation processes.

More recently, open labs have also been considered as a place suited for the organisation of the "*innovation commons*" (Potts, 2019). Innovation commons create the institutional conditions for an effective common pool of knowledge to accelerate the development of entrepreneurial opportunities (Allen and Potts, 2017). Innovation commons are considered as a model of innovation complementing the firm seen as a "nexus of contracts" as in Coase (1937, 1991) and Williamson (1990), and to the exploitation of knowledge assets, as in the resource-based view of the firm (Barney, 1991, 2001) and market mechanisms. Potts (2018) has explained that "innovation commons" are specifically relevant during the early stages of exploration during innovation processes. They reduce information access costs and knowledge articulation costs for entrepreneurs (Potts, 2018; 2019). Allen and Potts (2017) and Cohendet et al. (2021) directly mention open labs as illustrations of innovation commons even though they do not use this generic term: Allen and Potts (2017) refer to hackerspaces while Cohendet et al. (2021) analyse fab labs. Both explain that open labs make a direct contribution to install the dynamics of innovation commons by managing communities where people with a heterogeneous background work and live together.

Middle ground, Triple and Quadruple Helix, innovation commons: this book expands on these three conceptual bodies and investigates the conditions for open labs to develop a large pool of shared knowledge to encourage creativity and innovation. Even though the Triple and Quadruple Helix concepts emerged as a concept used to analyse macroeconomic phenomena and innovation-related public policies, the conceptual frameworks provide opportunities to discuss the interaction between micro, meso, and macro levels in the management of innovation and creativity.

Managerial approaches of open labs as agile organisations

The multifaceted contribution of open labs in the open innovation context directly relates to their internal ways of working and managerial originalities. Open labs cannot be considered as classical organisations, like companies, because they serve the dynamics of communities. Mérindol et al. (2021) call them "communities-based organisations" because the dynamics of communities explain their development. Because of the self-organisation schemes inherent in communities, the boundaries of the open labs are blurred and change over time. By nature, open labs therefore represent flexible organisations.

Explicit organisational and managerial challenges underlie operations in open labs when connecting together several "worlds" and contributing to the design of new forms of collaboration. This managerial perspective represents another originality of open labs as organisations. This book explains that such managerial specificities do not emerge at random, and that original managerial competencies are required to offer the boundary conditions for knowledge exchange and value creation identified by Goermar et al. (2021). If open labs represent boundary organisations or organisational intermediaries, the teams in charge of the development of open labs therefore represent boundary spanners in ecosystems.

They use tools and physical spaces that can be respectively considered as boundary objects and boundary spaces. Boundary objects offer the opportunity to mediate collaboration and generate the appropriate mechanisms for knowledge articulation during the emergence of new innovative solutions (Versailles and Mérindol, 2019). Boundary spaces offer the opportunity to work inside a neutral space suited to the emergence of unexpected encounters and to collaboration across cognitive and organisational paradigms (Micek, 2020; Champenois and Etzkowitz, 2018). Sarpong et al. (2017) point out that the challenge of the development of these boundary spaces is to change practices related to innovation. Open labs represent both a cognitive and physical space; Hussenot (2021) has already applied the concept of organisational fluidity to analyse them.

This book shows how the key components of open labs (physical spaces, community, and portfolio of services) evolve over time. It investigates how open labs represent an agile organisation and handle, enact, or combine, the three managerial dimensions of boundary spanners, boundary spaces, and boundary objects. This book investigates managerial challenges underlying their constant evolution to adapt to the needs of communities and ecosystems. Specific organisational and managerial challenges apply to open labs because of their links with communities that still require further investigation. However, this book will show that these managerial modalities must be proactively considered for open labs to perform their role and make an impact on ecosystems. The same holds for the business model of open labs, that still presents a series of open questions to generate sustainable and independent ventures. Open labs must still find effective solutions to break even and balance their mission, cost structure, and revenues. This book shows that the business model of open labs remains a traditional one, adhering to the rationales of the "old economy", even though it does also confront the challenge of handling the different variables supporting the analysis of the dynamics of communities. In traditional organisations, managerial processes aim at the production of specific "outputs"; in open labs, managerial processes aim at serving the community. The book will show that the sustainability of open labs depends entirely on the lifecycle of their communities.

Structure of the book

In the Foreword, Elise Tissier, Bpifrance Le Lab director, draws on perspectives about the open lab phenomenon and shows the relevance of this phenomenon for public policies.

This book is then divided into three parts and ten chapters. All chapters opt for a qualitative approach based on interviews, visits to open labs, and direct observation. In the chapters, field research elaborates on unique or multiple cases, investigating open labs located in France, Spain, Canada, and Asia.

The first part of this book investigates common features of open labs as agile organisations. Two chapters provide an analysis of their main strategic and organisational dimensions.

In Chapter 1, Valérie Mérindol and David W. Versailles offer a taxonomy of open labs based on key organisational attributes. The chapter zooms out from eight years of field research investigations in France and in Asia (funded by Bpifrance Le Lab and Innovation Factory or developed in partnership with the ANRT FutuRIS platform) and develops an analysis of communities hosted by open labs, their interaction with physical space and the evolutions of service portfolios. This taxonomy represents a tool for further research.

In Chapter 2, David W. Versailles investigates the rationales behind the elaboration of business models for the open labs, and perspectives for their sustainability. The chapter explains rationales for profitability and sustainability, in building links with strategic intentions. The chapter explains that these aspects are most often difficult to appraise, because open labs are often encapsulated into complex organisational and legal designs. The chapter explains that the business models of open labs instantiate rationales prevailing in the old economy, with a strong emphasis on occupancy rates, threshold effects, and the management of fixed costs. The main visible tension is between sustainability and scale, but the main distinctive assets lie in the dynamics of the community. This means that the root causes for sustainability relate to the dynamics of knowledge lifecycles and the preservation of strong ties inside the open lab communities, nurtured by a proactive management of the different components of their operations. Handling all these aspects at the same time proves to be a difficult equation that ultimately wraps up in the size of available working capital, and in the survival of communities. It is difficult to define *ex abrupto* the optimal size of a community, but field research already made it possible to point out that an open lab will find it difficult to create value beyond genuine consulting if its community dies.

The second part of the book investigates the multifaceted contribution of open labs as innovation intermediaries.

Chapter 3 specifically investigates open labs active in the domain of arts and culture. In this chapter, Nicolas Aubouin calls them "open art labs". He analyses how they create value with different configurations to serve the traditional functions of open labs in a renewed and transversal approach. The chapter focuses on three open art labs in France to describe the diversity of roles enacted by artists when they contribute or support innovation processes: explorers (promoters of a new vision creating bridges between different worlds), boundary spanners (Levina and Vaast, 2005) serving the articulation of different expertise, and co-producers of innovation (Imbert and Chauvet, 2013). The dialogue between artists and open art labs creates a fertile environment where the artists' contributions can take three contributions to make an impact on the innovation process: pollination, hybridisation, or pervasion. The chapter shows the issues at stake when creating value around artistic projects, or working with artists, and identifies expected managerial contributions to facilitate these activities.

Chapters 4 and 5 both work on healthcare ecosystems.

In Chapter 4, Alexandra Le Chaffotec and Valérie Mérindol focus on living labs in healthcare ecosystems. They show that living labs have three different contributions as open labs: energise healthcare innovation ecosystems by bringing together hospitals, private companies, and other institutions in reference to the user-centric approach of innovation; act as architects of innovation through the organisation of specific events (e.g., hackathons) enabling co-creation activities; or promote the emergence of communities of practice in enrolling end users into innovation processes. Beyond the rejuvenation of existing ecosystems, the chapter unveils the importance of the physical space to serve different options in the dynamics of communities. Building on the difference between their interstitial, user-friendly, or functional attributes, the chapter shows that different designs of the physical space and different managerial activities around it serve the dynamics of communities in original ways. Interstitial spaces foster the emergence of communities of innovation. User-friendly spaces support the development of communities of practice.

In Chapter 5, Luc Sirois and Karl-Emanuel Dionne investigate the temporal dynamics at play in innovation ecosystems. They analyse the tempo of events to nurture the dynamics of communities and of innovation projects. This chapter is based on a unique case study, Montréal-based Hacking Health network, where both authors are/have been "insiders". This chapter shows that events are essential tool orchestrators to be used in connection with innovation spaces to orchestrate the dynamics of ecosystems and handle the tempo of innovation. The authors explain that a space without events is like hardware without software, or bodies without souls. The authors claim that, in the absence of events, nurturing communities becomes impossible and open labs cannot deliver results. The authors show that the temporality of events feeds the epistemic contributions made by open labs: events are used to train, teach, create, and share knowledge at moments in time, therefore nurturing tacit and explicit knowledge inside innovation communities. The chapter shows the importance of synchronicity and temporality in the management of knowledge articulation processes in open labs, and in their interaction with innovation communities.

Chapter 6 by Luc Sirois, Octave Niamie, and Patrick Cohendet investigates the open lab "Communitech" in the Kitchener-Waterloo region, Canada, that act as an ecosystem of entrepreneurs. The authors show the various managerial processes in place to support the emergence of this open lab, and the importance of the culture of collaboration prevailing in that region to create a momentum around this open lab. The chapter shows how the Communitech open lab has progressively developed an associative model strengthening the sense of belonging for entrepreneurs, and a physical space reflecting the values of collaboration. The open lab was initially focused on start-ups, and gradually expanded to handle interactions with large companies. This development trajectory illustrates rationales for the focalisation of innovation-related activities around this open lab and the parallel development of the associated community of innovation. The chapter also shows how the agility of this open lab made it possible to handle specific shows in the ecosystem and become the "Silicon Valley of the North" leading to the formation of hundreds of technology start-ups, creating tens of thousands of jobs. The chapter explains the importance of managing and planning activities in and around this open lab, and it also identifies a long series of questions to address the sustainability of the model and ensure the resilience of the ecosystem after the pandemic crisis.

In Chapter 7, Ignasi Capdevila analyses the transposition of the open lab concept into rural ecosystems, most notably around the function of coworking spaces. The chapter provides an analysis of the role and of the management of open labs in rural environments and focuses on the development of new communities in this specific context. Capdevila investigates the Cowocat Rural case, coworking spaces located in rural Catalonia, Spain. He explains that the dynamics of communities tend to be limited to the physical space in urban environments, while it is expanded to territorial embeddedness in rural cases. Because of the low density of people in rural environments, the main challenge of the coworking spaces is to overcome the gap between external and internal communities, and the limitations incurred by the physical space hosting the coworking space. To ensure the development of communities, the chapter explains that coworking spaces need to deploy the management of the physical spaces beyond their boundaries, and to animate the dynamics of interactions outside them. The management must feed a virtuous circle of community development where activities and events hosted in the space progressively diffuse throughout the community, ensure its attractiveness, and diffuse to new members. While the emphasis always goes on internal collaborative dynamics for urban coworking, the chapter explains that the key to sustainability and success in rural coworking lies in the embeddedness of coworking practices into the local environment, with an explicit attention paid to reasons making the territory a focal point of attractivity in regional policies.

The last part of the book investigates how open labs influence the emergence of new governance models for innovation that were introduced earlier in this introduction. Three chapters contribute to this analysis. In Chapter 8, Olivier Irrmann introduces a link with the "middleground" concept. In Chapter 9, the authors working around Patrick Cohendet and Laurent Simon at MOSAIC research centre in HEC Montreal illustrate the link between open labs and the concept of innovation commons. In Chapter 10, Valérie Mérindol and David W. Versailles refer to the Triple and Quadruple Helix models of innovation. Even if they join different theoretical debates and illustrate different concepts, all three chapters stress show the importance of open labs in the emergence or in the facilitation of epistemic mechanisms in innovation processes.

Chapter 8 focuses on the emergence of communities of innovation in tertiary education and in the public service. Olivier Irrmann explains the bottom-up introduction of multi-disciplinarity in the educational system, and of design-based approaches in public administrations. He shows how local initiatives progressively percolated to the rest of their respective organisations in local ecosystems. In both cases, the processes started with an epistemic community (Cohendet et al., 2014) and a local physical space, to then gain leverage with projects and progressively build the "middleground". The cases explain the conditions for independence and interstitiality in these (constrained) environments. They show that the concept of "middleground" can be applied beyond the traditional frameworks of the management of creativity and that open labs play a prominent role in the transformation of the respective ecosystems.

Chapter 9 has been prepared by a large team of authors with a twin expertise as researchers and practitioners about the TransMedTech (iTMT) case, an open lab (living lab) installed in Montréal, Canada: Nathalie Tremblay, Patrick Cohendet, Geneviève Cyr, Margaux Manent, Laurent Simon, Marie-Pierre Faure, and Carl-Eric Aubin. Marie-Pierre Faure is currently Deputy Director in this open lab; Carl-Eric Aubin is the Founding Director, currently

Executive and Scientific Director. iTMT was one of the first open lab initiatives in Canada, with a focus on medical technologies and user-centric innovation processes. The chapter shows the sequences progressively building the community in interaction with "knowledge commons", but it stresses the need for the articulation of different "commons" to establish interdisciplinary boundary-crossing. The chapter shows how knowledge, innovation, social, and symbolic commons follow each other in a logical and temporal sequence around an open lab adhering to the rationales of a "middleground" and becoming an actual hub for innovation in medical technologies.

In Chapter 10, Valérie Mérindol and David W. Versailles explain that open labs have become the catalysts of many collaborations between the public and private actors, acting as boundary spaces supporting the development of the dynamics of Triple or Quadruple Helix innovation governance modes. They analyse cases in the fields of healthcare and smart cities to show the interplay between knowledge, consensus, and innovation spaces to build this catalyst role. They also show the necessity to install trust and legitimation mechanisms when these do not exist before open labs start their operations. The authors also explain the importance of appraising the contribution made by open labs from the "knowledge-based view" perspective of organisations, and the major importance of open lab managerial teams to facilitate the social learning cycles inside the open labs, and between the different contributors of interactions between the Triple or Quadruple Helixes.

The book concludes with an Afterword by Michel Ida, currently in charge of supervising and heading projects on societal impact of sciences and technologies at the French CEA (French Alternative Energies and Atomic Energy Agency), and formerly the Founding Vice president heading the open lab networks at CEA Tech. Michel Ida has been supporting, and contributing to, the newPIC chair initiatives on open labs since the beginning of our research in this area. In his Afterword, he shares his 20 years' experience in the domain. Michel Ida shows the importance of building meaning for the future to best anticipate issues in relation to the diffusion of technologies and innovation. He explains the fallacies following the resurgence of "magical thinking" (Levi-Strauss, 1966) and points out the main challenges incurred by the transition towards the new patterns of "sustainability centric" innovation.

References

Agogue, M., Ystrom, A., & Le Masson, P. (2013). Rethinking the role of intermediaries as an architect of collective exploration and creation of knowledge. *International Journal of Innovation Management*, *17*(2), 1350005-1–1350005-24.

Allen, D. W. E., & Potts, J. (2016). The origin of the entrepreneur and the role of the innovation commons. *SSRN Electronic Journal*, (November), 14 pages, article 2867850, Conference paper for the 86th annual meeting of the Southern Economic Association conference, Washington DC, Nov. 2016.

Amin, A., & Cohendet, P. (2004). *Architecture of knowledge: Firms, capabilities and communities*. New York: Oxford University Press.

Barney, J. B. (1991). Firm resources and sustained competitive advantage. *Journal of Management, 17*(1), 99–120.

Barney, J. B. (2001). Is the resource-based "view" a useful perspective for strategic management research? Yes. *Academy of Management Review, 26*(1), 41–56.

Bathelt, H., & Cohendet, P. (2014). The creation of knowledge: Local building, global accessing and economic development-toward an agenda. *Journal of Economic Geography, 14*(5), 1–14.

Bogers, M., Chesbrough, H., & Moedas, C. (2018). Open innovation: Research, practices, and policies. *California Management Review, 60*(2), 5–16.

Bouncken, R., & Aslam, M. M. (2019). Understanding knowledge exchange processes among diverse users of coworking-spaces. *Journal of Knowledge Management, 23*(10), 2067–2085.

Bouncken, R. B., & Reuschl, A. J. (2018). Coworking-spaces: How a phenomenon of the sharing economy builds a novel trend for the workplace and for entrepreneurship. *Review of Managerial Science, 12*(1), 317–334. https://doi.org/10.1007/s11846-016-0215-y.

Brown, J. (2017). Curating the "third place"? Coworking and the mediation of creativity. *Geoforum, 82*(April), 112–126.

Bürkner, H. J., & Lange, B. (2020). New geographies of work: Re-scaling micro-worlds. *European Spatial Research and Policy, 27*(1), 53–74. https://doi.org/10.18778/1231-1952.27.1.03.

Capdevila, I. (2015). Co-working spaces and the localised dynamics of innovation in Barcelona. *International Journal of Innovation Management, 19*(3), 1540004-1-28.

Carayannis, E. G., & Campbell, D. F. J. (2009). "Mode 3" and "quadruple helix": Toward a 21st century fractal innovation ecosystem. *International Journal of Technology Management, 46*(3/4), 201.

Carayannis, E. G., & Rakhmatullin, R. (2014). The quadruple/quintuple innovation helixes and smart specialisation strategies for sustainable and inclusive growth in Europe and beyond. *Journal of the Knowledge Economy, 5*(2), 212–239.

Champenois, C., & Etzkowitz, H. (2018). From boundary line to boundary space: The creation of hybrid organizations as a triple helix micro-foundation. *Technovation, 76–77*, 28–39.

Chesbrough, H. (2003). *Open innovation: The new imperative for creating and profiting from technology*. Boston, MA: Harward Business School Press.

Coase, R. H. (1937). The nature of the firm. *Economica, 4*(16), 386–405.

Coase, R. H. (1991). The nature of the firm: Origin, meaning, influence. In O. E. Williamson & S. G. Winter (Eds.), *The nature of the firm* (pp. 34–74). Oxford: Oxford University Press.

Cohendet, P., Grandadam, D., & Suire, R. (2021). Reconsidering the dynamics of local knowledge creation: Middlegrounds and local innovation commons in the case of FabLabs. *Zeitschrift fur Wirtschaftsgeographie/The Geman Journal of Economic Geography, 65*(1), 1–11.

Cohendet, P., Grandadam, D., Simon, L., & Capdevila, I. (2014). Epistemic communities, localization and the dynamics of knowledge creation. *Journal of Economic Geography, 14*(5), 929–954.

Etzkowitz, H., & Zhou, C. (2017). *The triple helix: University-industry-government and entrepreneurship*. London: Routledge.

Etzkowitz, H., & Leydesdorff, L. (2000). The dynamics of innovation: From national systems and 'mode 2' to a triple helix of university-industry-government relations. *Research Policy, 29*(2), 109–101.

Fab Lab. Studio. https://fablab.studio/histoire/dans-le-monde/, consulted in Sept. 2021.

Garrett, L., Spreitzer, G., & Bacevice, P. (2017). Co-constructing a sense of communities at work: The emergence of communities in coworking spaces. *Organization Studies, 38*(6), 821–842.

Goermar, L., Barwinski, R. W., Bouncken R. B., & Laudien, S. M. (2021). Co-creation in coworking-spaces: Boundary conditions of diversity. *Knowledge Management Research & Practice, 19*(1), 53–64.

Hargadon, A., & Sutton, R. I. (1997). Technology brokering and innovation in a product development firm. *Administrative Science Quarterly, 42*(4), 716–749.

Heraud, J. A. (2017). Science and innovation. In H. Bathelt, P. Cohendet, S. Henn, & L. Simon (Eds.), *The Elgar companion to innovation and knowledge creation* (pp. 56–74). London: Edward Elgar Publishing.

Howell, T., & Bingham, C. (2019). Coworking spaces: Working alone, together. *Kenan Institute of Private Enterprise, UNC Kenan-Flager Business School, The University of North Carolina at Chapel Hill, NC, USA.*

Howells, J. (2006). Intermediation and the role of intermediaries in innovation. *Research Policy, 35*(5), 715–728.

Hussenot, A. (2021). All for one, one for all! From events to organizational dynamics in fluid organization. *M@n@gement, 24*(2), 1–22.

Imbert, G., & Chauvet, V. (2013). Faire coproduire le client en conception innovante. Les quatre processus mobilisés par les sociétés de conseil en innovation. *Revue française de gestion, 39*(234), 167–183.

Jacobides, M. G., Cennamo, C., & Gawer, A. (2018). Towards a theory of ecosystems. *Strategic Management Journal, 39*(8), 2255–2276.

Jakonen, M., Kivinen, N., Salovaara, P., & Hirkman, P. (2017). Towards an economy of encounters? A critical study of affectual assemblages in coworking. *Scandinavian Journal of Management, 33*(4), 235–242.

Levina, N., & Vaast, E. (2005). The emergence of boundary spanning competence in practice: Implications for implementation and use of information systems. *MIS Quarterly, 29*(2), 335–363.

Levi-Strauss, C. (1966). *The savage mind.* Chicago, IL: University of Chicago Press.

Mérindol, V., Aubouin, N., & Capdevila, I. (2021). « Articuler confiance, hiérarchie et prix au sein des communautés : le cas des modes de coordination dans les open labs ». *Management International, 25*(HS), 184–205.

Mérindol, V., Bouquin, N. Versailles, D. W. Aubouin, N., Capdevila, I., Le Chaffotec, A., Chiovetta, A., & Voisin, Th. (2016). *Le Livre Blanc des Open Labs. Quelles pratiques? Quels changements en France?* Travaux du groupe d'experts co-animé par ANRT/FutuRIS et la chaire newPIC de PSB, Paris: ANRT et PSB (Mars).

Mérindol, V., & Versailles, D. W. (2017). *Créer et innover aujourd'hui en Île-de-France : le rôle des plateformes d'innovation.* Research funded by Bpifrance Le Hub, Innovation Factory et Paris&Co (with contributions by Ignasi Capdevila, Alexandra Le Chaffotec, Nicolas Aubouin, Marion Desnost et Marianne Cohen for field research). Paris: PSB.

Mérindol, V., & Versailles, D. W. (2019). *Créer et innover aujourd'hui en France et en Asie : Le rôle des plateformes d'innovation et des open labs d'entreprise.* (Contributions by Innovation is Everywhere for field research). Research Funded by Bpifrance Le Lab and Innovation Factory. Paris: PSB.

Mérindol, V., Versailles, D. W., Le Chaffotec, A., Aubouin, N., & Capdevila, I. (2018). *Créer et innover aujourd'hui en France : le rôle des plateformes d'innovation dans les écosystèmes régionaux*. Research funded by Bpifrance Le Lab and Innovation Factory (with contributions by Oceane Duyck and Salim Moulmaaz for field research). Paris: PSB.

Merkel, J. (2017). Coworking and innovation. In H. Bathelt, P. Cohendet, S. Henn, & L. Simon (Eds.), *The Elgar companion to innovation and knowledge creation* (pp. 570–588). London: Edward Elgar Publishing.

Micek, G. (2020). Studies of Proximity in coworking spaces: The basic conceptual challenges. *European Spatial Research and Policy, 27*(1), 9–35.

Ollila, S., & Elmquist, M. (2011). Managing open innovation: Exploring challenges at the interfaces of an open innovation arena. *Creativity and Innovation Management, 20*(4), 273–283.

Potts, J. (2018). Governing the innovation commons. *Journal of Institutional Economics, 14*(6), 1025–1047.

Potts, J. (2019). *Innovation commons: The origin of economic growth*. Oxford: Oxford University Press.

Ranga, M., & Etzkowitz, H. (2013). Triple helix systems: An analytical framework for innovation policy and practice in the knowledge society. *Industry and Higher Education, 27*(4), 237–262.

Roberts, J. (2017). Community, creativity and innovation. In H. Bathelt, P. Cohendet, S. Henn, & L. Simon (Eds.), *The Elgar companion to innovation and knowledge creation* (pp. 342–359). London: Edward Elgar Publishing.

Sarpong, D., AbdRazak, A., Alexander, E., & Meissner, D. (2017). Organizing practices of university, industry and government that facilitate (or impede) the transition to a hybrid triple helix model of innovation. *Technological Forecasting and Social Change, 123*, 142–152.

Schmidt, S., & Brinks, V. (2017). Open creative labs: Spatial settings at the intersection of communities and organizations. *Creativity and Innovation Management, 26*(3), 291–299.

Suire, R. (2019). Innovating by bricolage: How do firms diversify through knowledge interactions with FabLabs? *Regional Studies, 53*(7), 939–950.

Versailles, D. W., & Merindol, V. (2019). Boundary objects as the missing link in the orchestration of resources: An exploratory study of Dassault Aviation Mirage IV and Rafale programs. *Management International, 23*(4), 102–117.

West, J., & Lakhani, K. R. (2008). Getting clear about communities in open innovation. *Industry and Innovation, 15*(2), 223–231.

Williamson, O. E. (1990). A comparative of alternative approaches to economic organization. *Journal of Institutional and Theoretical Economics, 146*(1), 61–71.

Part 1

A taxonomy of open labs and their business models

1 Appraising the diversity of open labs with a taxonomy

Valérie Mérindol and David W. Versailles

In the open and digital innovation context, new spaces such as "third places", fab labs, makerspaces, hackerspaces, coworking spaces, and incubators arise and play a central role to encourage new work practices and to accelerate innovation processes (Merkel, 2017; Howell and Bingham, 2019). They contribute to the development of an entrepreneurial mindset and offer new opportunities to access and integrate a diversity of bodies of knowledge to develop creative projects (Bouncken and Reuschl, 2018). In this chapter, we propose to zoom out from the specificities of the labels used for "third spaces" (such as coworking spaces, fab labs, living labs, etc.).

A significant volume of academic publications has already acknowledged the variety of these spaces (e.g., Gandini, 2015; Capdevila, 2015) and their original contribution to the development of territorial ecosystems of innovation (Suire, 2019). These spaces play a central role at territorial level as parts of the "middleground"; they also contribute to build a new model of innovation management based on innovations commons (Cohendet et al., 2021; Potts, 2019). Other scholars provided an investigation of these spaces as instances of new organisational phenomena based on fluidity (Hussenot, 2021) or as community-based organisations (Mérindol, Aubouin, and Capdevila, 2021). However, it remains difficult to understand the common and distinctive features of these spaces despite their different names or labels. This increases the difficulty of comparing their respective impacts on the ecosystems and to appraise their evolution over time.

As a general label, we propose using the term "open lab" for these organisational designs because they have all inherited their respective cultural backgrounds from the world of makers, of open-source software communities and, more generally, of open innovation. All open labs are installed with a strategic intent, a "**mission**" or a "purpose", framing its expected contribution to ecosystems, and targeting reference stakeholders. In a commercial environment, the reference stakeholders would be coined as the "target client base" but the word "client" is misleading for open labs because most of them are not totally at ease with a commercial orientation. Beyond this strategic intent, open labs always share three components (Mérindol and Versailles, 2017):

DOI: 10.4324/9781003125587-3

- **A local community** that plays a central role in creativity processes. Open labs encourage the emergence of a sense of community (Garret et al., 2017; Spinuzzi et al., 2019).
- **A physical space** where serendipitous encounters and dense knowledge exchanges take place (Parrino, 2013; Jakonen et al., 2017) that most often transforms into a totem place.
- **A portfolio of services** aimed at residents and at other stakeholders of the ecosystem, such as the rent of seats in a coworking space, the access to an incubation programme, the access to prototyping tools, or the organisation of events (Mérindol and Versailles, 2017).

The taxonomy introduced in this chapter elaborates on these four components, and on key reference concepts explaining roles, functions, and mechanisms at work behind these communities, spaces, and services. The taxonomy proposed in this chapter looks for major drivers building categories of open labs, that are also suited to appraise their development trajectories.

Based on the investigation of 22 open labs located in France and Asia, this chapter has four sections. The first one investigates key dimensions of the open labs that are relevant to build the taxonomy. The second one illustrates how to use this taxonomy to characterise and compare open labs. The third section illustrates how to use the taxonomy to analyse their evolution. The last section concludes on the necessity to develop further examination of these key dimensions, and their modalities, in further academic projects to better understand the phenomenon of open labs.

1.1 Main dimensions in the taxonomy of open labs

The taxonomy of open labs elaborates on four dimensions: their mission, the space where they develop activities, the associated community(ies), and the portfolio of services delivered by open labs to residents or other stakeholders. Each aspect will be analysed and illustrated in a specific subsection.

1.1.1 The mission: business, social business, versus not-for-profit

The open labs in the sample are owned by entrepreneurs, or they are steered by a group of passionate people committed to doing things and resolving problems in a different way. Open labs are active on different topics, such as information technologies, creative industries, digital transformation, sustainability, or creative industries but this list is not limited and topics evolve very often to adapt to new sectors, or to break down the topics of social and environmental responsibility into subtopics worth specific investigations with specific stakeholders.

The main reference suited to differentiating open labs from each other remains the orientation of their mission. In this taxonomy that adheres to Yunus et al. (2010: 310) presentation, we propose to distinguish between three strategic orientations: business, social business, and not-for-profit. In the sample, 11 open labs are business oriented, eight are social-business oriented, and three are not-for-profit oriented.

We illustrate this point with three cases. Over the 1990–2018 period, NUMA was a business-oriented open lab located in Paris, dedicated to digital transformation. It attracted digital entrepreneurs, freelancers, large companies seeking help to manage digital transformation, and, also, local policymakers interested in new insight to develop projects in relation with smart cities. An open lab located in Malaysia and Singapore, Found8, is social-business oriented. It offers a user-friendly space to develop networking and connections; projects apply creativity to the development of social and sustainable businesses. Electrolab is established in the Paris region. This not-for-profit oriented open lab claims the values of hackers at the kernel of its strategy and aims at doing technology differently. Electrolab attracts people who have a passion for the world of makers and who adhere to the values promoted by hackers: students, jobless people, technology geeks, or engineers disillusioned by the way of working on technologies in large companies. Electrolab offers access, in particular, to prototyping tools.

1.1.2 The local community: the open lab's strategic asset

The literature in management science has emphasised the role of communities to support the dynamics of innovation and of creativity in an open-world context (Amin and Cohendet, 2004). Communities refer to the emergence of a common identity among people who share the same passion or practices. They relate to the development of strong ties that enhance virtuous knowledge exchanges. Communities are based on reflexive trust, informal relationships, and they are also very often associated with self-organisation processes in relation with the development of exchanges and of collaboration (Adler, 2001).

People join open labs because they share the same values, passion, and interests. They also seek to access to various resources for their entrepreneurial projects (coaching, tools for fast prototyping, flex office). However, in all cases in the sample, interactions within the respective communities represent the main motivation for remaining in the open lab. "We meet people who have different ways of thinking and break down barriers: it allows us to dare and work on things that our own social codes would have forbidden" (start-up, MakeSense Space). The same motivation is shared by the various "users" present in the open lab: start-uppers, freelancers, students, employees coming from large firms, artists, etc. "The community is the most attractive aspect of the open lab. We initially come for machines, but we stay here for the interaction with people" (engineer, Electrolab).

Communities represent the open labs' strategic asset. They help to bring entrepreneurs and freelancers out of isolation. They represent opportunities for the mutualisation of physical and cognitive resources. They trigger serendipitous encounters (Jakonen et al., 2017). Communities offer a social context in which openness, goodwill, and reciprocity represent the central mechanisms for coordination. Their involvement inside communities helps individuals to manage their emotions when they face setbacks in their innovation project or entrepreneurial venture. The sense of the community emerges progressively from interactions during events, from joint activities during challenges, from meetings taking place inside the open lab (Garret et al., 2017): the dynamics of communities have a direct logical link with the collocation of activities inside the open labs' spaces.

Beyond these common elements, communities active in open labs differ in two main aspects: the diversity of knowledge and the dynamics of collaborative practices.

This diversity of knowledge potentially shared and available in the open labs represents an important issue at stake. Communities offer unprecedented access to knowledge and expertise for entrepreneurs and small ventures (start-ups, SMEs, not-for-profit organisations, etc.). This diversity enriches creative processes thanks to the opportunity of articulating various perspectives about the same problems. For large companies, it also offers unprecedented opportunities to connect with a wide variety of cognitive resources during exploration projects.

In open labs, the diversity of knowledge is always associated with the diversity of projects (Spinuzzi, 2012). In the open labs of the sample, the diversity of knowledge materialises the variety of individual competencies and experiences present in activities and projects. It is triggered by social statuses, and/or by demographic issues, and/or by cultural aspects prevailing among residents and contributors to projects. These aspects are not exclusive from each other.

The diversity of projects enhances a diversity of interactions based on experience, good practices, and the sharing of useful connections among people. In open labs such as MakeSense, this aspect best instantiates how to take advantage of the diversity of knowledge shared inside the community because the community of start-uppers shows a relative homogeneity: young and French entrepreneurs, well educated (very often with master's degrees and business school diplomas), with favourable personal environments. However, they can better develop their projects in sharing relevant contacts, in targeting immediately the most suited grants and sharing returns on experience and best practices to handle the subsequent administrative processes, etc.

Other open labs offer a combination of different aspects. La Paillasse is a bio-hackerspace located in downtown Paris. The knowledge diversity of La Paillasse's community is characterised by the diversity of projects, by the wide variety of fields covered by residents, and by a relative diversity for residents' social statuses. Specialisation domains in La Paillasse cover big data, UX design, biology, sociology, digital technologies, electronics, arts, etc. Residents are employees of large companies or students and faculties with positions in Paris universities. Other residents are entrepreneurs or even retired people. In China, HAX Accelerator offers another illustration of the nature of the diversity of knowledge inside its community. This open lab is dedicated to the acceleration of entrepreneurial projects dedicated to hardware in relation to information technologies. Residents are mainly engineers and entrepreneurs, but they come from different parts of the world: the USA, Asia, Europe. The diversity of knowledge reflects the different categories of hardware developed in the projects, and the wide variety of residents' nationalities. Such levels of diversity offer the opportunity to take advantage of various perspectives about the world in the same place to rapidly develop new disruptive projects based on hardware.

Communities can be also characterised according to the modalities of collaboration. Three options prevail in open labs where residents can "work alone together", or they can interact in the framework of structured collaboration rules by the open labs themselves, or they can follow patterns of "emergent" or "spontaneous" collaboration with other residents. Again, these modalities for collaboration inside open labs are not exclusive from each other. However, in many cases, a dominant pattern emerges inside each community or each open lab.

The "*working alone together*" modality is present in all communities and all open labs (Merkel, 2017). For people inside an open lab, this modality means that the dynamics of collaboration are low while the dynamics of knowledge exchange might be conversely important, yet based on reciprocal relations only. In the open lab, residents access information, knowledge, advice, and critics that contribute to accelerating their individual projects. This modality represents the dominant pattern inside the HAX community in Shenzhen. Exchanges among various start-uppers offer opportunities to learn rapidly from similar experiences, thus accelerating individual projects. Thanks to such interactions, entrepreneurs learn how to organise a chain of production for their innovative solutions faster. Creatis is a French open lab located in downtown Paris and dedicated to cultural industries. There, "working alone together" also represents the dominant pattern. Entrepreneurs focus on their own projects. Exchanges among entrepreneurs offer opportunities to identify best practices, avoid errors and, also, share useful connections.

In the other modalities of collaboration, two or more people work together on the same project and pursue the same goal.

When collaboration is "*structured*" by the open lab, the managers of the open lab act as architects of collaborative and multi-partner projects among residents, or between residents and external stakeholders. Many instances of these projects relate to needs identified by a large company or a public policymaker. Two open labs match this dominant pattern for their communities. thecamp, located near Aix-en-Provence in the south of France, focuses on societal disruption in a turbulent world. The team managing the open lab runs multi-partner projects to help large companies change their culture and adapt their products. They coach projects with co-creation methods; team members are executives coming from different large firms, entrepreneurs, students, and artists. They support the development of creative ideas, and then follow up with prototyping and quick experimentation.

La Paillasse is another illustration of this collaboration pattern. Even if there is significant adhesion to the dynamics of "*working alone together*" in this open lab, La Paillasse stands out by the capacity of its team to organise multi-partner projects. It is, for instance, the case with the EPIDEMIUM project, an open science project jointly steered with Roche, the international pharmaceutical company. Rationales are easy to describe. Roche owns a huge amount of data on cancer and gave free access to La Paillasse's community through an anonymised digital platform. They can explore these data with longitudinal analyses and investigate new solutions to follow up patients during medical treatments. Some 225 community members with a variety of backgrounds registered to contribute to the EPIDEMIUM project: artists, biologists, engineers specialised in electronics or data science, students, designers, data scientists, and medical doctors. All of them freely organise in small groups for the investigations before sharing all their results on the digital platform and during seminars. The management of La Paillasse and Roche executives run a committee to follow scientific and ethical issues, and evaluate projects and results uploaded on the digital platform. At the end of each challenge, or of each round of activities, the most interesting projects are selected, and Roche offered to work with their promoters on the feasibility and exploration protocols to develop actual solutions for healthcare practitioners. Each year, Roche funded the development of five projects. Depending on the interest of La Paillasse's participants and their personal interest in going further with their project, Roche either took over the entire property of the projects or shared it with them.

In other cases, communities lead to forms of "*emergent collaboration*": collaboration among residents emerges progressively from day-to-day interactions and from the identification of common interests. Bel Air Camp and ICI Montreuil illustrate this modality. Bel Air Camp is located near Lyon (France's third urban agglomeration by size). This open lab hosts small businesses via long term contracts, offering private offices, the flexibility to grow in the facilities, and access to shared services (tools for prototyping, storage, canteen, and concierge support). Entrepreneurs have lots of opportunities for interaction via the shared services. Some of these ventures (either start-ups or SMEs) hosted at Bel Air Camp even decided to organise the selection process for interns together: communication of available positions with business and engineering schools, reception of applications, and interviews. In ICI Montreuil, in the Paris suburbs, emergent collaborations are frequent and relate to for-profit and not-for-profit projects. The primary focus in ICI Montreuil relates to craftsmen in the creative industries. However, the persons in charge of a non-profit organisation specialised in theatrical activities for disabled young people are also installed in ICI Montreuil coworking space. Thanks to everyday cohabitation, craftsmen started discussing with them about theatrical projects and decided to work together. The outcome was the fabrication of specific wooden tools used during the production of theatrical events, for staging, or for stage sets.

This second subsection shows that the dynamics of collaboration inside communities take several forms, characterised here under the labels "structured", "working alone together", or "emergent". In all cases, the open lab managers play a prominent role to enforce the rules emerging "bottom up" from the communities, or to organise their activities with "top-down" rules.

1.1.3 The physical space(s): the open lab's totem place and its implantation

Parrino (2013) and Bouncken and Aslam (2019) have identified that the physical context plays a crucial role in understanding the emergence of the open lab communities. Face-to-face interactions play a central role in developing strong ties based on relational trust. Oksanen and Ståhle (2013) also identified the key characteristics of physical spaces suited to enhancing the dynamics of collaboration in creative processes: the space must be reconfigurable, attractive, user-friendly, and collaboration- and communication-enabling. It should also reflect and embody the values promoted by the open lab.

In open labs, managers always spend a lot of time organising the physical space. They devote lots of effort to the design of the physical space because it is intended to ease the fluidity of exchanges and to increase opportunities for serendipitous encounters. The strategic function of the physical space is acknowledged by managers and residents. The space contributes to enhance creativity and to develop the sense of belonging to the community. "With the design of the space, we create conditions for porosity and make it possible to think and act out of the box. We must create the conditions for encounters and shifted perspectives" (a manager at NUMA).

> [*It is*] a place where activities take place, always with a lot of animation and interesting events. [...] There is a lot of sharing when we are in a creative process. [...] You meet people without having to do anything. The simple fact of having an office here confronts us with novelty, discovery, and unexpected encounters.
>
> (a resident in The Tank)

Beyond these common features, two aspects differentiate the open labs: the size of the space and the installation of a unique space versus a network of physical spaces.

Open labs exhibit significant size differences, thus impacting the nature of potential activities. Some open labs opt for a small space. As an illustration, The Tank, an open lab located in downtown Paris specialised in the social impact and media industries, has a 150 m² physical space. By its size, The Tank can only accommodate a small number of residents. Other open labs have much larger facilities, for instance Liberté Living Lab and ICI Montreuil. Their spaces have several wide floors, a total of 2,000 m² and 1,700 m² respectively. This allows for the adaptation of the configuration of the space for meetings and events, prototyping, or for the development of collaborative projects. Open labs are often organised with several floors that are more or less open by design, and more or less reconfigurable. At Liberté Living Lab, the higher the floor is in the building, the higher the level of privacy for residents and projects. The ground floor is reserved for public events. ICI Montreuil is organised with a floor dedicated to coworking activities, and another one for prototyping tools and machines for craftsmen. People specialised in creative industries are working every day in the coworking facilities while craftsmen spend their time on the other floor that looks rather like a specialised workshop. Both at Liberté Living Lab and ICI Montreuil, managers work hard to create the conditions of fluidity and of potential encounters between the different types of populations. At ICI Montreuil, they organise thematic seminars and social events generating opportunities to meet other (categories of) residents. In most open labs, the location of the canteen or the kitchen plays a major role in generating encounters. At ICI Montreuil, managers decided to locate it at the intersection between the two floors; at Liberté Living Lab, a kitchen open to the public sells street food to residents and inhabitants of the district.

At the other extreme, several open labs are characterised as mega spaces because they operate very large facilities. Station F in Paris, thecamp near Aix-en-Provence, or Euratechnologies in Lille (on the border between France and Belgium) perfectly illustrate these atypical places by the size of their spaces. In these cases, open labs do not host a single community but several activities, eventually several communities, in parallel. Euratechnologies operates on more than 10,000 m^2 and runs an incubation programme with several intakes of more than 100 start-ups per year. Some of these ventures stay on site for acceleration activities after the end of the incubation phase. Thanks to the size of the facilities, this open lab offers a critical mass of start-uppers on site and the eventual access to a community of experienced entrepreneurs. This generates a snowball effect that is in itself attractive enough for large companies such as IBM, General Electric, and consulting firms to locate new business units around the open lab, in a sort of campus. Station F located in downtown Paris, in close vicinity to two major train stations that are also transportation hubs. Station F hosts more than 200 start-ups in its own incubator and several incubators and accelerators operated by large companies or by public institutions on its premises, a grand total of more than 1,000 start-ups. In Station F and in Euratechnologies, facilities also host permanent offices or representatives of different public institutions supporting innovation, funding programmes, public tax services, or advisors in different legal and economic areas (e.g., intellectual property).

The second key aspect describing the physical space relates to the number of facilities for each open lab. Some open labs in the sample have a unique space: Station F, thecamp, Liberté Living Lab, La Paillasse. It is then easy to organise activities around the residents, and to nurture the community with the different activities hosted in these facilities. Other open labs distinguish themselves in operating a network of physical spaces. China Bee+++ operates spaces in five Chinese cities. The same holds for Hubba, operating five different spaces across

Thailand. In France, ICI Montreuil was the pilot project of the Make ICI network, that now runs a network of several spaces branded "ICI [town]" (for instance in Marseille, Nantes, etc.) that are spread over Metropolitan France, and each has a specific specialisation among craftsmen. Make ICI has the goal of mutualising parts of the resources installed in each open lab of the network, and of installing a social caring environment suited to exchanges between labs, communities, and members (including with facilities to host people from a specific community in another open lab for the duration of their use of the local machines for their project).

In some cases, open labs own and operate several spaces in different countries. This network of spaces contributes to the internationalisation of the open labs' activities, which is, for instance, the case with MakeSense and its local offices in seven countries (France, Lebanon, Peru, Philippines, Mexico, Senegal, and Ivory Coast), or HAX Hardware Accelerator (Shenzhen in China and San Francisco in California, USA). MakeSense offers similar services in each location, while HAX Hardware Accelerator has specialised its services in the maturation of the business model and the connections with investors in the United States, tools and connections for quick prototyping, technological maturation, and pre-industrialisation of new hardware solutions are located in China. Found8 also owns two spaces (Malaysia and Singapore), with the ambition to internationalise its activities and replicate its activities in other countries in Asia and hold strong positions with similar incubation services in each country.

The size of the space incurs specific economic challenges which will be extensively discussed in the chapter of this book dedicated to the business model of open labs. However, there is a specific issue with the geographic implantation of the space inside the ecosystem: open labs must be easily accessible to attract the expected variety of profiles and the volume of projects required to meet its economic sustainability. The main important drivers leading to decisions about an open lab's localisation are agglomeration effects, the vicinity of major economic actors and important resources, the proximity of transportation hubs, branding and marketing issues, and the dynamics of the real estate market. This last point must be considered as a cost driving the business model of open labs, that will turn into a strategic motivation in relation to the proactive rejuvenation of industrial wastelands (as with Make ICI, ICI Montreuil, or Euratechnologies) or as an opportunity because of low rents in districts under transformation (this typically applies to Paris central and Eastern areas where most Paris-based open labs locate).

Last, but not least, the physical design and the implantation of the open space shall convey a specific image, consubstantial to its name and its (social and societal) impact. Beyond short-term effects and marketing issues, the space itself embodies the values it promotes, and eventually becomes a sort of totem impossible to separate from the open lab's brand. This is the case with most open labs considered in the sample. The most extreme cases here are thecamp, with its futuristic architecture in the emblematic landscape of

Provence, or MakeSense and ICI Montreuil where the space design and furniture are based on second-hand materials and wooden pallets recycled/upcycled from equipment used by residents and by the management of the open lab. At MakeSense, the design of the space itself immediately gives a sense of a makerspace because raw wood prevails everywhere. Open labs adhering to the "makers" or sustainability philosophy have it obviously easy to show off their adherence to these orientations in producing their own furniture or in recycling/upcycling as much as possible. However, it is always important that open labs pay careful attention to organise a close match between their equipment and their strategic orientation. All details are important for facilities to become a sort of totem place embodying both the implantation and the equipment to match the lab's strategic values.

1.1.4 The portfolio of services: a multi-sided approach

All open labs offer a portfolio of services. The list of the most frequent services covers several standard options: access to prototyping tools, incubation and/or acceleration programmes, organisation of events in relation to innovation and creativity or technology, hackathons, consulting, and coaching for large companies or public institutions (most often for creativity sessions or in relation to digital transformation). Almost all open labs have installed coworking spaces, but they implement different strategies for the renting of seats, depending on their business models, maturity, and ecosystem. At the beginning of their activities or when they face cash/revenue issues, open labs rent seats in the coworking space without much selectivity; they progressively focus on start-ups and other innovation-related ventures when they can afford this selectivity. Original services also emerge sometimes. Paris Saclay Hardware Accelerator (located in the south of the Paris region) and HAX Hardware Accelerator (in Asia) offer services specialised in the pre-industrialisation and industrialisation phases: they propose dedicated tools and machines (and the associated assistance and expertise) to build a supply chain and work on hardware solutions. China Bee+++ offers the rent of private offices and other mutualised services such as a fitness room, or the access to physiotherapy. Bel Air Camp accepts start-ups and SMEs during their ramp-up phase. Lots of open labs reserve their services for the start-ups selected in their incubation or investment programmes. Found8 works in that way: it invests at early stage in start-ups with huge social impact and does not host entrepreneurial ventures outside its investment portfolio.

To differentiate between the portfolios of services offered by open labs, our taxonomy uses two variables: the degree of specialisation in the services portfolio (or its "width"), and the role of the open lab's community(ies) in the delivery of these services.

The first variable is easy to characterise, and the portfolio of services is either large or focused, with a limited or large number of services, whatever the thematic specialisation(s). The expansion of the number of specialisations does not add to the number of services, even if it eventually expands the volume of activities. Two labs based in the Paris region illustrate the notion of a "focused" portfolio of services. Electrolab provides mutualised prototyping tools and the associated training on the use of machines and prototyping tools. Station F defines itself as an "academy of entrepreneurship" and focuses on delivering all the services required by start-ups at different stages, either with its in-house team or via partner companies hosting their thematic incubation and acceleration programmes in Station F facilities. Hospitality services (catering services and a business hotel located nearby) have now been added to the portfolio of services, while the access to specialised law firms and to offices representing French administration and public services supporting innovation were always offered to start-ups hosted in the programmes. Station F, Euratechnologies, Liberté Living Lab, and most other open labs, cover all aspects related to the creation of a new company.

Other open labs propose a wider portfolio of services. MakeSense in Paris and Found8 in Asia illustrate this strategy. Found8 has an early-stage investment function in start-ups, offers coworking spaces that are not related to its incubation programmes, offers coaching for freelancers and entrepreneurs on creativity issues, facilitates events such as hackathons and, also, runs an incubation programme. MakeSense also develops a similar width in its portfolio of services, but it does not perform any investment function. MakeSense also offers specific services for intrapreneurs coming from large companies. In the domain MakeSense also offers coaching in relation to creativity projects or the support to complex innovative projects run by large companies, or by education and public institutions. Last but not least, it offers access to a showroom for craft products developed in their network of stakeholders, or experiments developed by start-ups in MakeSense programmes worldwide.

All open labs mentioned in the two previous paragraphs follow different strategies with respect to their thematic specialisations. It adds to their complexity when they accept work on several topics in parallel, as with Station F, even though incubation services there are separated into different "modules" executed by different large companies hosted in their facilities. It may also simplify their activities when selecting a very precise topic, or a small list of precise topics. MakeSense opted for a thematic specialisation around social entrepreneurship. The tables in this chapter and the annexes to this book provide more instances of specialisations for all open labs listed in the chapter.

The second and more original variable relates to the operation of service delivery in the open labs, and to the role of communities in these operations. This taxonomy introduces a twofold relationship. First, service delivery is either reserved for start-ups and residents, or open to external "clients" (for instance, large companies or public institutions). Second, service delivery is either performed by the open lab's team only, or it mobilises the community of residents when working for other residents only, or for external "clients". This twofold relationship explains the very different options met in the field when visiting open labs and analysing their strategies.

Services delivered to the communities present in open labs are most often associated with the management of flex office or coworking spaces, with the access to, and the use of, tools for prototyping activities (with the associated training programme), and incubation programmes. Traditional services offered by open labs listed earlier are often offered to communities. As an illustration, La Fabrique, an open lab located in Strasbourg, operates as a makerspace, and focuses its activities on prototyping activities for residents. In Asia, HAX Hardware Accelerator also specialises its services on acceleration for start-ups hosted in their programme.

Another option develops services for external stakeholders (e.g., large companies) jointly delivered by the open lab's team and the community of residents. Liberté Living Lab illustrates such rationales. It manages complex innovation projects for large companies and public institutions. The open lab's team brings experts from its community in the projects and the creative processes, or staffs them to deliver specific consulting services. In such specific cases, the open lab appoints residents of its community for their contributions or uses their contributions to compensate for the fees that they should pay for being hosted in the facilities. In the sample, several open labs deliver services to large companies and public institutions but the residents' implication in such service delivery is not always expected. At NUMA and MakeSense, for instance, such consulting services are only delivered by the lab's employees and managers. At thecamp, specific categories of residents (the "hivers") are present in these activities, as parts of their "compensation packages".

Digital Village offers an explicit illustration of the combination of services provided for, and with, the community. Digital Village is an open lab located in downtown Paris and dedicated to freelancers specialising in web-related activities. This open lab reserves the access to flex offices for freelancers in residence. Through its start-ups' studio, it also provides an access to up-to-date innovation in digital technologies and cooperation with top engineering schools active in the same domain. To attract clients, the managers of Digital Village developed the brand and the visibility of the open lab. Potential contracts with clients then flow in thanks to the lab's management, that in turn presents these requests to the members of the community. Those who are interested and have time volunteer to participate in the project and contribute to the execution of projects. Digital Village therefore organises itself like a sort of agency representing its community, offering a platform for the elaboration of contracts and for the management of invoices with the external clients.

In a nutshell, the issue of the portfolio of services is not difficult to handle, but it is necessary to discriminate between the strategic specialisation of the open lab and the width of the portfolio, and the role of the community. Combining these aspects with the rationales underlying the dynamics of the local community and the role (and design) of the physical space makes it possible to develop an analysis of trajectories available with open lab ventures.

1.2 Using the taxonomy to identify similarities and differences

This section applies the different dimensions discussed in Section 1.1 to the sample of open labs presented in this chapter. To wrap up, Table 1.1 lists the different dimensions of the taxonomy and the associated sub-dimensions and their modalities. It also introduces the different codes used in Table 1.2 to characterise the different open labs. Some of these modalities are exclusive, others are not. Other modalities sometimes lead to the investigation of complementarities.

The different dimensions of the taxonomy resonate with each other.

An easy illustration lies in the fact that the mission of the open lab impacts the composition of the community and the nature of the services delivered to

Table 1.1 Dimensions of the taxonomy and illustrations

Dimensions	*Sub-variables*	*Modalities*	*Codes*
Nature of the mission	N/A	Business oriented	BUS
		Social-business oriented	SBUS
		Not-for-profit oriented	NFP
Dynamics of the community	Diversity of knowledge in the community	Correlated with the specialisation of projects	PROJ
		Specialised in various fields and competences	SPEC
		Based on specific demographic or sociological aspects	D/SOC
	Nature of collaboration	"Working alone together"	WAT
		Structured by the open lab	S
		Emergent with day-to-day interactions and common interests	E
Physical space	Size of the space	Small labs	Small
		Large labs	Large
		Mega platforms	Mega
	Number of spaces	Unique space	Unique
		Network of spaces	Network
Portfolio of services	Targets	Internal residents	Res
		External stakeholders	Ext
		Both internal and external	Res+Ext
	Operations	Open lab management only	Manag
		Management with contributions by residents	Manag+Res
	Specialisation	Specialised in topics	S/Topics
		Specialised in support to entrepreneurs	S/Entr
		Diversified to address external stakeholders' needs (e.g., consulting and support to creativity, support to intrapreneurs)	Diversified

Table 1.2 Characterisation of the open labs of the sample with the taxonomy

Open labs	*Mission*	*Community*		*Space*		*Services*		
		Diversity	*Collaboration*	*Size*	*Number*	*Specialisation*	*Targets*	*Operations*
ICI Montreuil	SBUS	PROJ + SPEC	WAT + E	Large	Unique	S/Topics	Res+ext	Manag + Res
Make ICI	SBUS	PROJ + SPEC	WAT + E	N/A	Network	N/A	N/A	N/A
NUMA	BUS	PROJ + SPEC	WAT + S	Large	Unique	Diversified	Res+Ext	Manag
thecamp	SBUS	PROJ + SPEC + D/SOC	S	Mega	Unique	Diversified	Ext	Manag
Found8	SBUS	PROJ + SPEC	WAT + E	Small	Network	S/Topics	Res+Ex	Manag + Res
Bel Air Camp	BUS	PROJ + SPEC	WAT + E	Mega	Unique	S/Topics	Res	Manag
Electrolab	NFP	PROJ + SPEC + D/SOC	WAT + E	Large	Unique	S/Topics	Res	Manag
MakeSense	SBUS	PROJ	WAT + E	Small	Network	Diversified	Res+ext	Manag
Liberté Living Lab	SBUS	PROJ + SPEC + D/SOC	WAT + S	Large	Unique	Diversified	Res+Ext	Manag + Res
Creatis	BUS	PROJ	WAT	Small	Unique	S/Entr	Res	Manag
Digital Village	BUS	PROJ	WAT + S	Small	Network	S/Topic	Res+Ext	Manag + Res
La Paillasse	NFP to BUS	PROJ + SPEC + D/SOC	WAT + S	Large	Unique	S/Topic	Res+ext	Manag + Res
China Bee+++	BUS	PROJ + SPEC	WAT	Large	Network	S/Topics	Res	Manag
HAX Accelerator	BUS	PROJ + D/SOC	WAT	Large	Network	S/Entr	Res	Manag
La Fabrique	NFP	PROJ + SPEC + D/SOC	WAT+ E	Large	Unique	S/Topics	Res	Manag
HUBBA	SBUS	PROJ	WAT	Large	Unique	Diversified	Res+Ext	Manag
Station F	BUS	PROJ + SPEC	WAT	Mega	Unique	S/Entr	Res	Manag
Euratechnologies	BUS	PROJ	WAT	Mega	Unique	S/Entr.	Res	Manag
The Tank	SBUS	PROJ	WAT	Small	Unique	S/Topics	Res+Ext	Manag + Res
Paris Saclay Accelerator	BUS	PROJ	WAT	Large	Unique	S/Topics	Res	Manag
The Corner	BUS	PROJ	WAT	Small	Unique	S/Entr	Res+Ext	Manag
Usine IO	BUS	PROJ+SPEC	WAT	Large	Unique	S/Topics	Res+Ext	Manag

residents and community members. The three orientations of the open labs (business, social business, not-for-profit) directly influence the residents' activities and what they are looking for when joining an open lab. Several open labs make it easy to illustrate how to use the taxonomy for the analysis.

NUMA is a business-oriented open lab. Over the years 2013–2019, NUMA was running an accelerator for start-ups, providing coaching activities for the digital transformations of large companies, and delivering project management services to facilitate complex collaborative projects in relation to creativity or innovation. Its business orientation was prescribing its activities and ways of working.

Found8 is an open lab with a social business mission: it mainly attracts social-business entrepreneurs, freelancers, students, and technological geeks interested in the resolution of problems with important social impacts.

As a not-for-profit open lab, Electrolab never sought to develop a large portfolio of services. A limited number of start-ups gravitated around Electrolab to access its various tools, machines, and expertise; as per its mission, this open lab never wanted to monetise its service.

The specifications of the open labs' spaces impact the size and dynamics of the communities. All open labs generate a series of formal and informal relationships between residents, or between residents and stakeholders, but their nature varies with the size of the space. In mega spaces such as Station F or Euratechnologies, residents work together in small groups which can turn into communities. In a mega space, public institutions also have formal offices and thus a more stable and personal interaction with residents. Mega spaces do not only work with one single creative community; they tend to encourage the emergence of a constellation of communities and foster more variated interactions with innovation ecosystems. In fact, this variety generates agglomeration effects at a much greater scale and the mega spaces therefore tend to act as catalysts in the dynamics of innovation ecosystems.

Open labs operating a network of different spaces logically serve several communities that "geographically" match each implantation. Community effects emerge at a local level, but they do not exist at network level unless the management of the open lab develops specific efforts to build these interactions and promote connections between spaces: digital and face-to-face interactions, collaborative projects, and/or joint events progressively contribute to the installation, or expansion, of community effects. The management of the open lab plays a key role in facilitating the development of strong ties (in local communities) and weak ties (interactions between local communities across the network).

Table 1.2 characterises the open labs considered in this chapter against the different aspects of the taxonomy. The other aspects of these open labs are documented in the annexes of this book. The taxonomy offers the possibility to compare open labs according to their various dimensions. The only line in the table that does not describe an actual open lab is Make ICI, because this structure rather represents a "network coordination head" for several open labs mimicking ICI Montreuil's key success factors in different local ecosystems. This is the reason why the table records "N/A" for "not applicable" in some cells, even though the different open labs affiliated with Make ICI all replicate the ICI Montreuil model.

The taxonomy and the data documented in Table 1.2 are used to compare open labs in crossing the characteristics of the physical spaces (size, uniqueness) for each strategic intent ("social business" or "business" orientation). It would be too complex to present all open labs on the same diagram, therefore Figure 1.1 compares "social business-oriented" open labs versus Figure 1.2 which compares "business-oriented" open labs. The figures help develop comparative case studies. As an illustration, when the challenge deals with the dynamics and the development of communities, it is relevant to compare the influences of variables underlying the interplay between the physical spaces and the mission of the open labs. The diagrams do not present the "not-for-profit" orientation as our sample is too small in this respect. The diagrams position the open labs vis-à-vis each other. The diagrams do not refer to any sort of quantitative ratings, and only reflect ordinal qualifications around the two axes. The

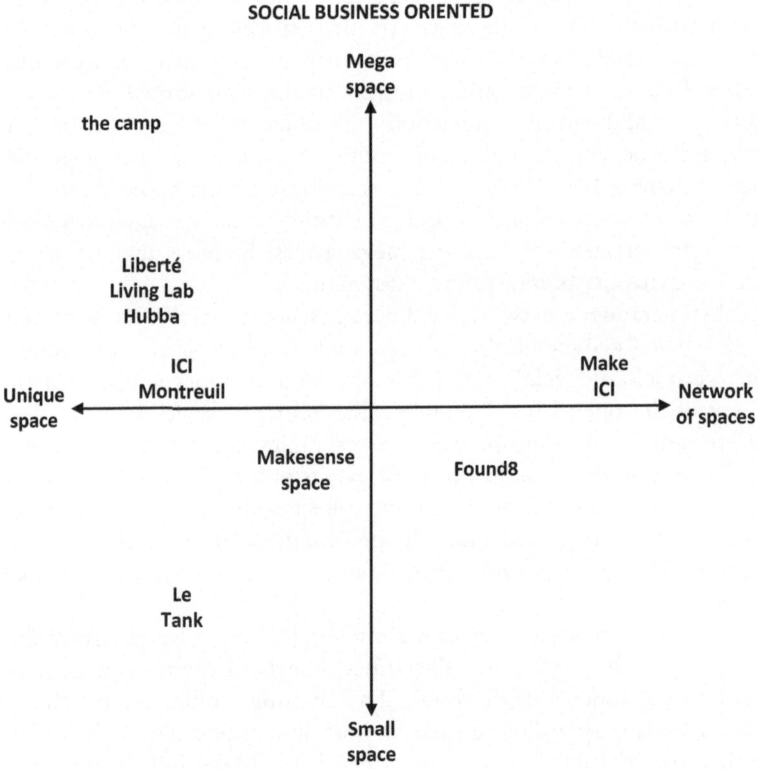

Figure 1.1 Social-business-oriented open labs.

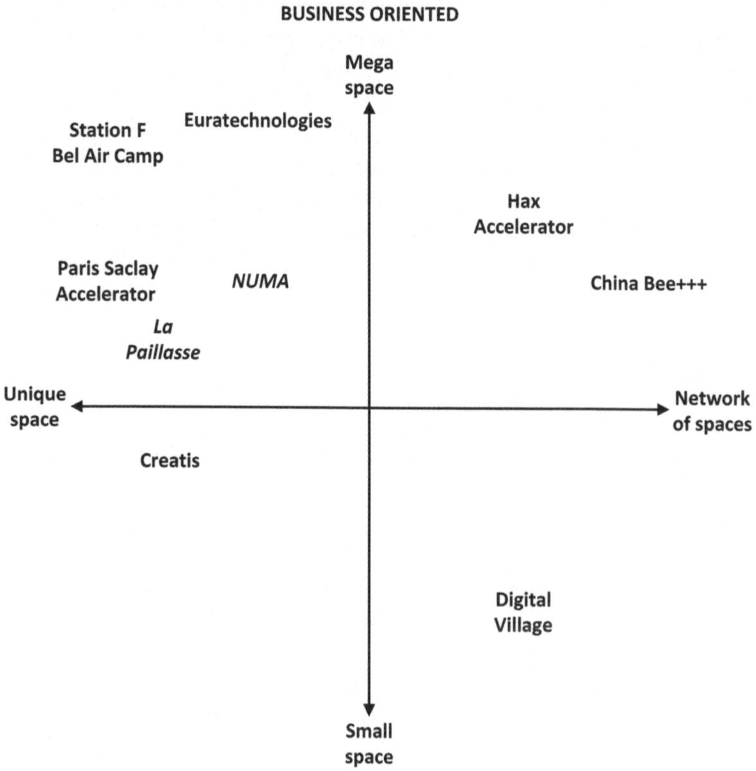

Figure 1.2 Business-oriented open labs.

open labs positioned in the vicinity of the axes are the ones that demonstrated a tendency to evolve their strategies.

In Figure 1.1, MakeSense space represents the main facility (located in Paris) for the international MakeSense community; for a long time, it was even its sole physical space. This strategy has evolved with opportunities to develop several sites in the Paris area in line with its original mission statement, and the creation of a network of sites in different countries worldwide. However, there is no deliberate strategy towards the creation of a network of open labs in France. This is not the case with Make ICI, which was designed as a sort of "network coordination head", and ICI Montreuil as a sort of prototype in the network. This is the reason why ICI Montreuil is located on the left side of the diagram while Make ICI is located on the far right and slightly higher than ICI Montreuil because the elaboration of a network is now one of the most important features in the strategy of this open lab.

Now other comments on Figure 1.2. NUMA and La Paillasse are located above the horizontal axis because both work with large facilities (three-storey buildings) in downtown Paris, as compared to Digital Village that can only host a maximum of 30 coworkers. However, facilities at NUMA and La Paillasse do not compare in size with mega open labs such as Euratechnologies (five campuses in Lille metropolitan area, the largest of which being the 80,000 m^2 Leblan-Lafont facilities in downtown Lille), Station F (facilities large enough to host 1,000 start-ups in downtown Paris, with a co-living space for 600 entrepreneurs, café, grocery shop, laundromat, restaurant, and all supporting services), or thecamp. These mega open labs are organised as comprehensive ecosystems to host residences and projects. thecamp offers a 170-seat amphitheatre, an ephemeral structure (170 seats and 200 standing places) to host events during the winter months in the nearby pine forest when the outdoor theatre (400 seats) is not available, 24 modular rooms to host meetings and projects (four to 50 seats each), a virtual reality room (five seats), a 120 m^2 fab lab dedicated to prototyping and creativity (operated by the Make ICI network), hospitality facilities with 161 rooms and a total capacity of 229 beds, a bar (50 seating spots), and a restaurant with two rooms and a large terrace. In addition, thecamp makes available to residents a 25 × 6 m swimming pool, a fully equipped 150 m^2 fitness centre, and several sports grounds. Most facilities dedicated to "users" of this mega open lab are designed with glass walls open to the Sainte-Victoire Mountain painted by Cezanne, to enjoy this typical Provence scenery even when the mistral wind blows hard and cold. This explains why thecamp stands out in the diagram with its "unique" and "mega" position.

It is also possible also to use data available in Table 1.2 in referring to other variables, such as the existence of the community, and compare, for instance, the dynamics of knowledge exchange inside communities in reference to collaborative behaviours.

1.3 Using the taxonomy to appraise evolution trajectories

Open labs represent organic structures. They constantly adapt their services and physical spaces to the needs of the community and of the ecosystem. Open labs show entrepreneurial capabilities and demonstrate their ability to constantly capture new opportunities of development. In our sample, many open labs have changed ("pivoted") since 2017. In some cases, the changes are so important that they lead huge transformations.

The taxonomy of open labs offers the opportunity to investigate these dynamics by identifying stabilities and evolutions for their main dimensions. The main changes relate to the portfolio of the services of the open labs and to the number of the physical spaces.

The diversification in the portfolio of services is easy to observe. Many open labs have extended their portfolio of activities to meet the ecosystem's needs, especially when addressing those emanating from large companies and policymakers. Such services are really similar to consulting activities to match processes in place in businesses for interaction with suppliers, or they comply with public tendering procedures. ICI Montreuil and NUMA perfectly illustrate this evolution. In these two open labs, initial services focus on the community's needs. Both NUMA and ICI Montreuil organise events to improve connections between their residents and boost interactions with external experts on specialised topics relevant for the community; they also run an open office with coworking seats reserved for residents, meeting their strategic targets (start-ups in the case of NUMA, craftsmen in ICI Montreuil). NUMA also provides incubation services while ICI Montreuil offers prototyping machines also available to produce a short series of products. However, many large companies seek the expertise available in these open labs to develop new innovative capabilities. At the beginning of the 2000s, large companies asked NUMA for support in the management of their digital transformation and the acquisition of new creative skills. In the early 2010s, large companies asked ICI Montreuil to develop new services based on the combination of expertise available, with various craftsmen hosted in the facilities. Both NUMA and ICI Montreuil then promoted a portfolio of services specifically dedicated to large companies. In parallel, ICI Montreuil diversified its portfolio of activities to support local policymakers in the rejuvenation of industrial wastelands in partnership with communities of craftsmen.

Many open labs provide incubation services because entrepreneurs represent a growing part of their communities. Liberté Living Lab, MakeSense, or The Tank have developed this service. Following the same logic, ICI Montreuil installed a partnership with Creatis for an incubator called "Made in France", dedicated to artistic and craft projects with local implantation. Initially dedicated to social entrepreneurs, La Ruche decided to extend its incubation service to refugees. This initiative contributes to modify the composition of its community and enrich it with new profiles. Some open labs only target entrepreneurs with a mature project. They tend to adapt their service portfolio to match these specific needs and expectations. In other cases, like Station F, open labs develop hospitality facilities reserved for the entrepreneurs hosted in the incubation programme, to ease the development of their projects during their residence. This is typically important in places like Paris where housing is expensive and difficult to find.

Many open labs progressively promote a network of physical spaces. They opt for diversification at local, national, or international levels. The expansion of their activities either goes through the replication of the same logic with other physical spaces located in other cities, or with the installation of a network of complementary activities, thus adjusting both dimensions at the same time (service portfolio and networked spaces). NUMA illustrates the first option: during the 2010s, it replicated its activities in India, Spain, and Russia. Digital Village also extended activities in Strasbourg (the main city in the north-east of France, on the border with Germany), with a new space for freelancers specialised in web activities. Make ICI and The Tank follow the second pattern and take advantage of local opportunities in different areas. They extend their portfolio of activities (incubation services, consulting activities for large companies) and also create new spaces in several French cities. The accumulation of changes along these dimensions can radically transform the open labs. First, by impacting the dynamics of communities with a wider connection range and/or greater knowledge diversity inside the community. Second, by assigning a more strategic role to local communities, as with Make ICI and ICI Montreuil. The open lab is not only a way to pool resources for craftsmen. Its local community and the available network also generate leverage to rejuvenate industrial wastelands in city centres.

The sample shows the relative stability of the strategic orientation installed by open labs. The business, social business, and not-for-profit orientations remain unchanged in the majority of the cases. La Paillasse represents the only case where the initial mission was reshaped. The mission was initially not-for-profit oriented, with the goal of doing science differently, in reference to the hackers' values and to open science. Upon the request of large companies and to meet the expectations of a growing number of entrepreneurs inside the community, La Paillasse now provides new services for large companies, and supports their acquisition of new modes of management for explorative projects based on multidisciplinary competences. These consulting activities imply an expansion in the size of the open lab team and, also, diversification of the composition of the La Paillasse community with many employees from large companies joining to learn new practices to manage science project. La Paillasse today lies at the intersection between the not-for-profit and business orientations. Many members of the community remain attached to the hackers' values, but some residents now deal with business concerns.

The adaptability of open labs represents one of their key characteristics. Open labs are flexible enough to adapt their activities to the constantly evolving landscape of innovation and creativity management. The taxonomy presented in these pages offers the opportunity to understand their evolutionary trajectories. It also offers the opportunity to identify when open labs turn into traditional firms, or into some sort of consulting businesses. This is typically the evolution illustrated by NUMA, which stopped being an open lab around the year 2020 (see Chapter 2). NUMA was created in the early 1990s and legitimately considered itself a first mover in the world of open labs in France. Its service portfolio evolved over time. In the 2000s and early 2010s, NUMA was considered as one of the most famous incubators/accelerators for digital start-ups in Paris. The analysis of its business model shows that the volume of incubated/accelerated start-ups never made it possible to break even with the sale of stakes in start-ups and with the available working capital. Activities (and revenues) then shifted to consulting activities for large established companies, still without meeting the volume of cash required to cover recurring costs. The situation worsened for NUMA when other open labs emerged in Paris with strong brands and business models adhering to other rationales. After several attempts at innovation management-related consulting, NUMA then pivoted to executive education. It does not host a community anymore.

1.4 Conclusion and perspectives for further research

The chapter provides a definition of open labs and goes beyond the specificities of their contributions as spaces of innovation, and of labels or main missions (fab labs, makerspaces, incubators, and coworking spaces). The chapter identifies common features. The chapter coins the term "open lab" and investigates the interaction between key dimensions: a community, a physical space, and a portfolio of services. The taxonomy proposed in this chapter makes it possible to compare the open labs, to analyse their strategies, and to understand their trajectories of development. This taxonomy also serves the analysis of the business model of open labs, most notably with the investigation of their portfolio of services, and the preparation of the analysis of operational expenditures. The rationales explained in Chapter 2 to discuss the sustainability of the business model of open labs use all variables presented in the taxonomy.

The taxonomy proposed in this chapter is also useful for policymaking supporting the development of open labs. Open labs represent a new category of organisations encouraging innovation at territorial level. Their contributions are original. They foster the connections among people coming from various institutional spheres. They enhance serendipitous processes, and they also encourage the development of new creative projects. By characterising their mission, their portfolio of services, and the diversity of activities inside their communities, this taxonomy represents an effective tool to support conceptual analysis of open labs, or help (local) policymakers in the design of innovation-related policies taking advantage of open labs.

References

Adler, P. S. (2001). Market, hierarchy, and trust: The knowledge economy and the future of capitalism. *Organization Science*, *12*(2), 215–234.

Amin, A., & Cohendet, P. (2004). *Architectures of knowledge: Firms, capabilities, and communities*. Oxford: Oxford University Press.

Bouncken, R., and Aslam, M. M. (2019). Understanding knowledge exchange processes among diverse users of coworking-spaces. *Journal of Knowledge Management*, *23*(10), 2067–2085.

Bouncken, R. B., and Reuschl, A. J. (2018). Coworking-spaces: How a phenomenon of the sharing economy builds a novel trend for the workplace and for entrepreneurship. *Review of Managerial Science*, *12*(1), 317–334. https://doi.org/10.1007/s11846-016-0215-y.

Capdevila, I. (2015). Co-working spaces and the localised dynamics of innovation in Barcelona. *International Journal of Innovation Management*, *19*(3), 154004.

Cohendet, P., Grandadam, D., & Suire, R. (2021). Reconsidering the dynamics of local knowledge creation: Middlegrounds and local innovation commons in the case of FabLabs. *Zeitschrift fur Wirtschaftsgeographie/The Geman Journal of Economic Geography*, *65*(1), 1–11.

Gandini, A. (2015). The rise of coworking spaces: A literature review. *Ephemera: Theory and Politics in Organization*, *15*(1), 193–205.

Garrett, L., Spreitzer, G., and Bacevice, P. (2017). Co-constructing a sense of communities at work: The emergence of communities in coworking spaces. *Organization Studies*, *38*(6), 821–842.

Howell, T., and Bingham, C. (2019). Coworking spaces : Working alone, together. *Kenan Institute of Private Enterprise*. Kenan Institute Working Paper. Retrieved from https://www.kenaninstitute.unc.edu/wp-content/uploads/2019/04/Coworking_04042019.pdf.

Hussenot, A. (2021). All for one, one for all! From events to organizational dynamics in fluid organization. *M@n@gement*, *24*(2), 1–22.

Jakonen, M., Kivinen, N., Salovaara, P., and Hirkman, P. (2017). Towards an economy of encounters? A critical study of affectual assemblages in coworking. *Scandinavian Journal of Management*, *33*(4), 235–242.

Merindol, V., Aubouin, N., and Capdevila, I. (2021). « Combiner confiance résiliente et réflexive, hiérarchie formelle et prix au sein des communautés : Le cas des open labs », *Management International*, sous presse.

Merindol, V., & Versailles, D. W. (2017). Créer et Innover aujourd'hui en Ile de France : Le rôle des plateformes d'innovation, rapport d'étude commandée par Bpifrance LE LAB et Innovation Factory, NewpiC Chair, PSB. http://www.newpic.fr/02proj2016openlabsidf.html.

Merkel, J. (2017). Coworking and innovation. In H. Bathelt, P. Cohendet, S. Henn, & L. Simon (Eds.), *The Elgar companion to innovation and knowledge creation* (pp. 750–597). London: Edward Elgar Publishing.

Oksanen, K., & Ståhle, P. (2013). Physical environment as a source for innovation: Investigating the attributes of innovative space. *Journal of Knowledge Management*, *17*(6), 815–827. https://doi.org/10.1108/JKM-04-2013-0136

Parrino, L. (2013). Coworking: Assessing the role of proximity in knowledge exchange. *Knowledge Management Research and Practice*, *13*(3), 261–271.

Potts, J. (2019). *Innovation commons: The origin of economic growth*. Oxford: Oxford University Press.

Spinuzzi, C. (2012). Working alone, together: Coworking as emergent collaborative activity. *Journal of Business and Technical Communication*, *26*(4), 399–441.

Spinuzzi, C., Bodrožić, Z., Scaratti, G., and Ivaldi, S. (2019). "Coworking is about community": But what is "community" in coworking? *Journal of Business and Technical Communication*, *33*(2), 112–140.

Suire, R. (2019). Innovating by bricolage: How do firms diversify through knowledge interactions with FabLabs? *Regional Studies*, *53*(7), 939–950.

Yunus, M., Moingeon, B., and Lehmann-Ortega, L. (2010). Building social business models: Lessons from the Grameen experience. *Long Range Planning*, *43*(2–3), 308–325.

2 The business model of open labs

Sustainability at the intersection between scale and community life cycles

David W. Versailles

This chapter investigates open labs from the specific perspective of their business models. It explores the main important variables explaining sustainable strategies in open labs, and the links between these variables. It elaborates on a sample of 15 open labs located in France. These open labs have been presented in Chapter 1. In France, whatever their orientation, as fab labs, living labs, makerspaces, coworking spaces, etc., open labs have attracted major attention in public policies on innovation and entrepreneurship. They were most often supported with public subventions during the initial years of the open lab movement. A transition started around the years 2005–2010 when public subventions started to progressively vanish. However, today, they need business models that do not only rely on subventions anymore. Open labs are still looking for a sustainable and profitable business model. The challenge is important because it combines the complexities of setting up and sustaining a business model, and of "managing" the community it hosts. As identified in Chapter 1, communities represent the key asset of open labs. As an intangible asset, communities remain very complex to manage.

This chapter considers a business model as a series of interactions between services, demand, operations, and capital (or financial resources) (Johnson et al., 2008). In line with Budler et al. (2021) it uses business models as units of analysis to approach open labs. This chapter looks for business model design options for open labs and points out successful drivers. It investigates how sustainability and profitability relate to the scale effects present in the business models used by open labs. How is it possible to manage the tension between scale, sustainability, and profitability? Applied to business models, it also requires a link with strategic intents, and a link with the dynamics of communities.

The chapter shows, first, that the business model of open labs is framed by fixed costs; it responds to references borrowed from the "old economy" to serve innovation and creativity projects. Second, it shows that the identification of the proper scale to launch activities represents a key strategic question and, paraphrasing the popular adage, that the sky is not the limit: the business model of open labs cannot grow *ad infinitum*. Open labs develop and expand until strong ties characterising communities dissolve into weak ties, thus disincentivising clients and users.

DOI: 10.4324/9781003125587-4

In this chapter, Section 2.1 focuses on the tension between profitability, sustainability, and strategic intentions, with direct links to legal statuses. Section 2.2 analyses the tensions between sustainability, profitability, and scale. Section 2.3 discusses the links between sustainability, scale, and the dynamics of knowledge interaction in communities. Section 2.4 concludes.

2.1 The tension between profitability, sustainability, and strategic intention

Open labs follow a strategic intention, and they often represent entrepreneurial ventures in themselves. It is important to make a difference between the strategic intention underlying the open lab and its business model, and the orientations of communities hosted in the open lab previously discussed in Chapter 1: "business oriented", "social business-oriented", versus "not-for-profit-oriented" communities. This first subsection proposes a theoretical model to expand the key characteristics of the open labs proposed in Chapter 1 and start the analysis of their business models with considerations about profitability. The subsequent subsections will give data on the cases, and show the tensions between strategic intentions, organisational designs, and communities.

2.1.1 Sustainability and "social" returns

The characteristics of communities and the taxonomy of open labs proposed in Chapter 1 suggest that open lab initiatives are not always driven by profitability concerns. The analysis of business models in open labs is located at the intersection between several approaches because they mix considerations of innovation, sustainability, and social returns. However, jumping too fast to conclusions remains dangerous: McMullen and Bergman (2017) illustrated, for instance, that a "prosocial motivation" shall be cautiously considered. Even if there is no automatic link with some sort of social activism (Santos, 2012), the reference to social entrepreneurship is relevant for open labs because "social entrepreneurs" often "prioritize social returns on investments" over "business orientations" (Elkington and Hartigan, 2008).

This chapter proposes an investigation of the links between sustainability, profitability, and scale in reference to four categories of business models. Definitions are borrowed from Elkington and Hartigan (2008: Chapter 1) illuminated by the discussion of social entrepreneurship by Short et al. (2009).

- **Leveraged non-profit ventures** meet needs ignored by current market mechanisms and businesses. Organisations following this pattern usually deliver a public good to those who cannot otherwise access it. They generate cash from charity and public funding. As identified by Short et al. (2009: 170), such ventures solely foster "social improvements in society".
- **Hybrid non-profit ventures** blend non-profit and revenue-generating strategies: in these businesses, "the notion of making (and reinvesting) a

profit is not out of the question" (Elkington and Hartigan, 2008: Chapter 1). Working at the intersection between public/non-profit management and entrepreneurship, social entrepreneurship contributes to new social value creation (Short et al., 2009: 170).

- **Social business ventures** are for-profit entities focused on social missions, most often with a focus on transformational, social, and/or environmental change. Firms installed under this strategic intention combine social and financial returns (Elkington and Hartigan, 2008: Chapter 1). When informing the intersection between social issues and entrepreneurship, social entrepreneurship "is concerned with value creation that impacts the relationship between organizations and societal stakeholders" (Short et al., 2009: 171).
- **For-profit ventures** are usually to caricature under the sole idea of maximising profit for shareholders, but the perspective of creating value for all its constituencies is consistent with the model. This perspective is in line with considerations about corporate social responsibility (Carroll, 1991) and the approaches considering the necessity to organise a co-development between firms, their ecosystems, and their stakeholders. Working at the intersection between entrepreneurship, social issues, and public/non-profit management, "the distinctiveness of social entrepreneurship lies in using practices and processes that are unique to entrepreneurship to achieve aims that are distinctively social, regardless of the presence or absence of a profit motive" (Short et al., 2009: 171–172).

It is pure tautology to point out that the issue of profitability does not apply to "leveraged non-profit" ventures. This business model option was typical for pioneer open labs in their initial phase, whatever their activities. The sustainability of these projects either depends on the recurrence of public subventions and charity, or on the transformation into another category of venture.

In the three other options for open labs, the discussion of the open lab's sustainability therefore mandates an analysis of the interplay between profitability and scale. The analysis of open labs must then overcome lots of ambiguities because entrepreneurial motivations (Gruber and McMillan, 2017) promoting the open lab projects, the dynamics of communities hosted in the open labs, and the various innovation or start-up projects developed inside them may eventually diverge. These dimensions are all intertwined yet require some level of alignment. Consistency issues must be addressed to define the appropriate "target customers" and define the potential for sustainability in the business model.

2.1.2 Encapsulation of open labs into complex organisational designs

This chapter now expands the initial analysis based on communities to business model modalities. Table 2.1 describes such cross analysis.

Table 2.1 Strategic intent in the business model versus dynamics of communities

Communities	*Not-for-profit communities*	*Social-business-oriented communities*	*For-profit oriented communities*
Business models			
Non-profit business model	• Electrolab • La Fabrique		• TUBA
Hybrid non-profit business models	• La Paillasse	• La Ruche • MakeSense	
Social business ventures		• ICI Montreuil, Make ICI • Liberté Living Lab	
For-profit ventures			• NUMA • Bel Air Camp • Station F • Make It Marseille • Garages XYZ, Paris Saclay Accelerator • Usine IO • The Corner

Despite the apparent diagonal presented in Table 2.1, it remains difficult to generate deterministic links between the orientation of communities and the taxonomies applicable to open labs, because open labs address two series of ambiguities at the same time. First, they often encapsulate their activities into wider organisational designs. Second, they tend to separate activities and organise with several legal (and commercial) entities to best align legal (and commercial) forms, strategic intentions, orientations of communities, and revenue sources.

The first ambiguity relates to the encapsulation of the open labs into wider organisational designs. At first glance, NUMA or Station F are easy to identify as "for-profit" ventures articulating with "for-profit" communities of residents. The reality is not so simple. NUMA, thecamp, or Station F illustrate the first difficulty at using the taxonomy adapted from Elkington and Hartigan (2008).

NUMA was among the pioneer open lab initiatives in France; it emerged in the 2010s from different initiatives in the Sentier district in downtown Paris, France. NUMA claims itself as the first French open innovation programme in France, active in the domain of digitalisation and digital transformation. NUMA emerged as in independent venture, the capital of which was fuelled by a crowdfunding process. The position of NUMA in Table 2.1 makes sense when limiting the analysis to NUMA itself. However, the historic governance and current recapitalisation of NUMA show the ambiguity of coining NUMA as a for-profit initiative, as the French mutual insurance company MAIF was always present as one of their shareholders. MAIF has now become NUMA's sole owner. MAIF strongly adheres to the "hybrid non-profit" models as a mutual company. NUMA's "for-profit" model was therefore always encapsulated into a "non-profit" venture: its "for-profit" business model aims at sustainability, fuelling MAIF "hybrid non-profit" strategy.

Station F (in Paris), thecamp (in the Aix-en-Provence area), and Bel Air Camp (in Lyon) follow similar rationales: they represent subsidiaries of profitable and successful holdings respectively owned by a venture capitalist and entrepreneur, by the heirs of a successful entrepreneur, and by a real estate developer. These projects open the way for an appraisal of profitability that does not solely locate at the level of the open lab itself, but also at the level of the respective parent companies. Station F and Bel Air Camp illustrate a "for-profit" orientation in the open labs aligned with a "for-profit" orientation of the communities present in the open labs, and a link with "for-profit" holdings. In thecamp, the business model adheres to "social business" rationales: thecamp hosts a "for-profit" community (inside the incubator or the makerspace) at the same time, and artists showcasing a wide diversity of motivations while also delivering services to large companies with a "for-profit" motivation. thecamp belongs to the "for-profit" holding owned by Frederic Chevalier's heirs (Chevalier was the founder of thecamp and was a successful entrepreneur in the tech and communication industry) that assigned a "social business" logic to the open lab. In these three cases, the sustainability of the open lab project depends on capital injections operated by the parent company and on revenues. Sustainability and profitability cannot be assessed at the sole level of the open lab because they fuel the analysis of profitability operated by the holding companies on which they depend.

The second ambiguity deals with the separation of open lab activities into different legal entities that are each associated with specific rationales as regards business models and communities. When emerging with "non-profit" business models, open labs most often selected the status of non-profit organisations (in French: "association sans but lucratif"). This status generates constraints because it frames service delivery and forbids competition against commercial firms, and complexifies the management of charging and taxation schemes (for instance: supplies are purchased "VAT included"). Open labs have therefore transitioned towards other legal statuses, consistent with their community members' motivations. The difficulty for open labs is in the necessity of adapting their strategic intention and the orientation of their business model to the categories of communities. In France, the standard status operated today by open labs has shifted towards the simplified stock company (in French: "société par action simplifiée", SAS). This legal form is usually selected because of its adaptability to different corporate social purposes, and of its simplified governance framed in the statuses of the company (and not by the general commercial law). SASs are subject to standard regulations for value added tax (VAT) and corporate tax.

As a recent evolution, several open labs such as Liberté Living Lab have also added the reference to the "ESS" special label in their legal form and the associated governance. "ESS" is the French acronym for "social and solidarity economics". It was introduced in the French legal system under law #2014-856 published on July 31, 2014. Such companies should adhere to a strategic intent that is not profit-sharing based; they shall enact a "democratic" and "participative" governance; the major parts of benefits shall be used to sustain or develop companies' activities. The "ESS" label has similarities with the "B Corp" brand granted for "social and environmental performance" to for-profit companies in the UK or in several American states. The "ESS" label may apply to the SAS legal status, as for any other sort of legal form of company listed in the commercial law.

Several open labs have decided to split their activities among several legal entities complementing each other, and best adapt their respective strategic purposes and their business models to the stakeholders' expectations. This is typically the case with MakeSense. MakeSense, the kernel of the network, is a non-profit organisation covering the coworking space and the organisation of events. Commonsense, a branch created in 2012 for consulting services to corporate clients, is a stock company (SAS) that pays its fees to the MakeSense non-profit organisation. SenseSchool is an educational entity specialising in programmes with "positive impact". It has the status of a non-profit organisation. SenseCube is an incubator created in 2014 in Paris; another incubator is located in Brussels and operates under Belgian commercial laws.

The analysis of profitability, sustainability, and strategic intentions therefore requires a cautious characterisation covering several dimensions where business model modalities, legal forms, and strategic intentions should be appraised in conjunction with the dynamics of communities hosted in the open labs. There is no simple one-to-one and deterministic link between these variables.

2.2 The tension between sustainability and scale

The analysis of business models elaborates on an always-growing body of literature in management science. Even though the concept of business model often lacks clarity (Foss and Saebi, 2017), recent contributions (e.g., Budler et al., 2021) point out that it represents a viable unit of analysis for developing research. The chapter adheres to the evolutionary and domain-based perspective opened by Foss and Saebi (2018). After some theoretical comments, different subsections explore related diversification in open labs, revenue sources, strategies to cover fixed costs, and scale issues.

The reference framework for the analysis of business models goes back to four key aspects identified by Johnson, Christensen, and Kagermann (2008): a value proposition for customers, a profit formula, key resources, and key processes. It is easy to replace the words "profit formula" with "balanced budget" or "positive EBITDA" (earnings before interests, taxes, depreciation, and amortisation) to still refer to their seminal definition. As pointed out by Johnson, Christensen, and Kagermann (2008), "profit formula" and "business model" are not interchangeable words for how one makes a margin or a profit, they only represent one piece of strategy.

Figure 2.1 provides concrete guidance to work on these aspects. At the centre of the business model lies the service model, with the portfolio of services and products offered to create a customer value proposition. The portfolio of services (and products) should match the needs characterising target customers. Traditional questions elaborate on the "job to be done", on the details of value creation, and on value capture. When discussing these aspects, it is also necessary to understand the rationales for rarity of products/services, non-imitability, and non-substitutability of such a portfolio of activities, thus building the eventual robustness of the business model (as codified by Barney and Hesterly, 2005).

Consistently with the resource-based view (RBV) approach (Barney, 1991) and as identified by Johnson, Christensen, and Kagermann (2008), the operation model investigates key resources and key processes. Service (and product) delivery frame operations, shaping activities beyond "production". An important issue relates to the discussion of technical skills, collective competences, and individual competencies required to materialise the business model. All these aspects translate into costs, to be computed under the difference between fixed and variable costs, direct and indirect costs, wrapped up on the diagram (see Figure 2.1) with the acronym OPEX ("operation expenditures"). The computation of economies of scale and threshold effects emerges automatically in the analysis.

On the other side of the diagram is the revenue model. At first glance, this part of the diagram seems easy to investigate because revenues are the multiplication of prices by volumes of sales. Going into deeper detail, this model should characterise target clients, their purchasing power, and the volume of products or services they are able to afford and willing to pay for. Target clients sometimes only order specific categories of services if they are free of charge, or eventually exchanged against other services. This analysis leads to the elaboration of prices and to the definition of pricing schemes, impacting the service portfolio retroactively to match affordability. Volumes should be appraised

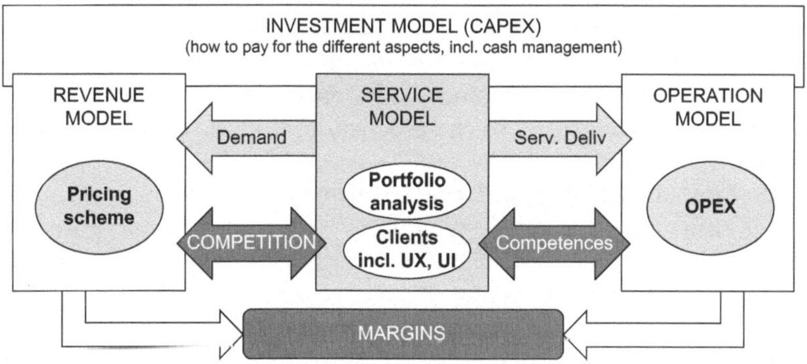

Figure 2.1 Generic tool for business model analysis.

in terms of market size, purchase frequency, ancillary sales, etc. Options for related diversification and economies of scope automatically emerge.

The elaboration of a business model requires a "big picture" leading to some sort of "design to cost" process where a 360° vision of market opportunities translates into the adaptation of the service model and of the operation model. These elements are finally framed by the analysis of the investment model, that covers capital expenditure (CAPEX), the management of cash, and of working capital.

The words "business model innovation" apply to any novel way to think about the components of Figure 2.1, or their combination. As in Foss and Saebi (2018), this chapter reserves these words for evolutions that have a strategic and systemic dimension. Readers can refer to Andreini, Bettinelli, Foss, and Mismetti (2021) for a more precise review of such an approach.

The following subsections now discuss (related) diversification, the management of revenues and fixed costs, and scale issues.

2.2.1 Service model: related diversification prevails

The first easy distinction about open labs deals with the dichotomy between "makers" and "thinkers" that provides structure to the service portfolio. Open labs in the "makers" movement support prototyping and deliver services in relation to the fabrication of actual prototypes, or a short series of products. In the "thinkers" orientation, open labs facilitate (or operate) creativity and innovation projects with end users or professionals. They also organise events, meetings, hackathons, or any sort of innovation-related event for their clients.

Zooming out from the different instances of open labs investigated over the years, it is possible to identify five main categories of services. These services have already been presented in Chapter 1.

- **Support to prototyping and other "maker" activities:** this part is directly linked with the presence of tools for prototyping. Beyond the traditional 3D printing machine used as an emblematic illustration, prototyping should adapt to different categories of technological content, and of technological intensity. Associated services should not only cover operations, but also training and support to the operation of machines, and the establishment of a safe environment. The operation of prototyping tools may also eventually expand to the production of short series of products for SMEs, craftsmen, or artists.
- **Incubation (all phases) and acceleration:** the interaction with start-ups represents a major trend for open labs, with specialised services providing a nest for entrepreneurial ventures. Services must adapt to the different stages of entrepreneurial ventures, and facilitate both the interaction between start-ups, and the interaction with mentors, experts, or coaches.
- **Support to creativity projects:** this aspect may either represent the kernel of the open lab activities, as in living labs, or one of the "big" topics

when such services are delivered for start-ups and external clients; it may also be totally absent from open labs. There is no general rule. It is usually only organised and facilitated by the internal staff; start-uppers sometimes contribute to the facilitation of such activities as a sort of counterpart to their presence (as in the Hive programme at thecamp, or at Liberté Living Lab).

- **Space rental (events):** the open lab's physical space has become emblematic of the dynamics of open innovation (Mérindol, Versailles and Le Chaffotec, 2022). Facilities are most often designed with the twin purpose of reconfiguration and flexibility for the open lab's activities. Open labs subsequently have the opportunity to offer their facilities for rent without putting other activities at risk. thecamp in Aix-en-Provence has set up a futuristic space in the neighbourhood of the famous Sainte-Victoire Mountain painted by Cezanne. Liberté Living Lab in the Sentier district in the centre of Paris also organised its facilities with an open area dedicated to events on the ground floor, while other floors are reserved for innovation projects and incubation.
- **Workstations in a coworking space:** most open labs are organised with an open space and workstations reserved for incubated start-ups or creativity teams. In cases of overcapacity, and when available, workstations are not used for incubated start-ups or innovation projects, it is possible for the open labs to propose those facilities for rent.

As an illustration, Table 2.2 shows how different open labs adhering to the "makers" versus "thinkers" orientations articulate the components of their service portfolios.

Open labs have expanded their service portfolios over time because they all were constrained to generate cash from several sources. Support to prototyping, incubation (acceleration), and support to creativity projects represented core aspects in the initial portfolios. The most ancient initiatives expected a specialisation on a single specific action. The diversification of activities occurred because services in relation to incubation or acceleration, and support to prototyping, were never generating enough cash to cover the costs, and working capital was tight. Open labs have therefore been pushed towards a strategy enabling a presence at all stages on this continuum of services to follow the money and target clients able to afford the services mentioned. Such adaptation was easy to perform in the "thinkers" open labs while it was much more difficult to introduce in the "makers" labs because of the size of investments. This typically explains why the focus on prototyping soon developed into the production of small series of products for craftsmen or SMEs (except at ICI Montreuil which planned it upfront), or why incubation services for start-ups soon evolved into the facilitation of intrapreneurship programmes for large companies. The same holds for space rental, that was not intended to represent a direct source of revenues, but soon pragmatically evolved into a specific service.

Table 2.2 Illustrations of service portfolios in various French open labs

Open lab	*ICI Montreuil*	*Liberté Living Lab*	*Usine IO*	*TUBA*	*The Corner*	*thecamp*	*La Ruche*	*Make Sense*	*Bel Air Camp*	*La Paillasse*
"MAKERS"	☒		☒							☒
"THINKERS"		☒			☒	☒	☒	☒	☒	
Machinery and support to prototyping	☒		☒			☒			☒	☒
Incubation and acceleration	☒				☒	☒	☒	☒		☒
Support to creativity projects	☒	☒		☒	☒	☒	☒	☒		
Space rental (events)	☒	☒	☒		☒	☒	☒	☒	☒	☒
Space rental (workstations)	☒	☒	☒	☒	☒		☒	☒	☒	☒

2.2.2 Revenue model: pragmatism prevails

Field research made it possible to identify four main categories of clients purchasing services from open labs: public institutions, large established companies expecting support for intrapreneurship, innovation and creativity projects, SMEs, and start-ups. Chapter 1 already comments on these clients. This subsection explains the evolution of revenues earned from each category of client.

Most open labs delivering services for start-ups have built their business models on the perspective of exchanging incubation or acceleration services for shares without "charging" start-ups in a direct way, but this compensation cannot generate cash on the short run. A blurry perspective does exist to generate cash in the future when reselling this stake, but this does not solve the need for cash in the short term. None of the open labs observed during field research ever made it possible to generate this balance so far: the ramp-up towards possibilities for sales of these stakes remains too slow and demands very high levels of working capital, notwithstanding the uncertainty incurred by the evolution of the start-up value. This is the reason why open labs need either to develop related diversification and find new sources of revenue, or to increase working capital through recapitalisation.

The case of NUMA illustrates this difficulty. NUMA is famous in France for raising €1 million on the YesWeCrowd crowdfunding platform in 43 days in 2013, from 330 contributors. NUMA enacted a "first mover" advantage in France over the years 2011–2017, with a special focus on acceleration. The initial plan was to accelerate start-ups in exchange for a stake, but NUMA's CEO explained in an interview to the French newspaper *Les Echos* published on January 25, 2019, that this business model had never been viable. Services delivered to large companies represented c.80% of NUMA's total revenues in 2013–2014. Public records for NUMA for the fiscal year 2018 show a total revenue short of €8.5 million, at a total loss of €3 million. Attempts for related diversification investigated two directions: international expansion to scale up NUMA activities and scout for start-ups in other countries, and thematic extension to be present on new technological or societal challenges. NUMA communication identifies 13 different acceleration programmes active in 2015. The press also lists the installation of only eight international offices out of the 19 planned ones, because NUMA was short of cash. The race between revenues and costs was lost; the break-even point was not met. Between 2017 and 2022, NUMA consequently operated two strategic moves. It refocused first on consulting services. Today, it only operates an educational mission for executives. NUMA has abandoned all other services.

Other options exist to charge start-ups. Several open labs, such as the incubators affiliated with Paris&Co, explained that they expect incubated start-ups to pay "hosting" fees and compensate for services in returning to them part of the grants distributed by Bpifrance, the French public agency supporting innovation projects and start-ups. Most open labs do not directly charge for incubation or acceleration services. They charge for "workstations" used by start-ups in their coworking space and in this way cover a small part of their costs. Sometimes, as in Liberté Living Labs, they also host experts or other practitioners relevant for their other activities in the coworking space, for a monthly price different from the one charged to start-ups. "Workstations" represent a source of revenue in most open labs. They associate with an "objective" pricing scheme, fair to all community members because it is proportional to the number of employees in the start-ups or in the project. It most often represents a second best used by open labs after they realise that they are short of cash and that they need to cover fixed costs.

This chapter already mentioned that the open lab movement has been initially supported by public subventions in France. Public grants to open labs were initially justified for their "generic contribution" to ecosystems. Public budgets have not disappeared from the list of revenue sources, but institutions now expect explicit service delivery against their budget injections, and often run public calls for proposals where open labs compete for these budgets against other ventures. ICI Montreuil and the network of open labs headed by Make ICI have specialised into such calls where they support (or operate) public policies aiming at regional development, or at the rejuvenation of industrial brownfield sites. TUBA, in Lyon, has adapted to the same dynamics and now wins public calls for tenders to operate open innovation and living-lab activities in relation to societal transformation.

The elaboration of services dedicated to large established companies and SMEs emerged when open labs managed to successfully exhibit their originality at running user-centric innovation, creativity, open innovation, or intrapreneurship projects, and their distinctive advantage against large companies. Liberté Living Lab illustrates how a strong partnership with one of the main French banks generated a win-win solution; it also offered a mid-term sustainable solution for the open lab business model. The case for services delivered to SMEs and craftsmen is different and is mainly articulated by open labs operating in the "makers" orientation: clients purchase services in relation to the use of machines, including the associated supplies and expertise, to complement their own activity in a sort of B2B interaction. For these clients, using the open lab's services makes it possible to improve their own B2C service portfolio without the need to equip with machines. This option typically describes the rationales for ICI Montreuil.

Illustrating the dynamics of revenues for open labs is not easy because actual data are not easily available. During a public event organised in their facilities in 2016, Usine IO explained the division of revenues: 20% from large companies, 20% from SMEs using their machines, and 60% from start-ups. thecamp recently disclosed details about its division of revenues, after a significant reorganisation. They publicly discussed a recapitalisation process after the shutdown of activities in late 2020 and the first quarter of 2021 due to the Covid pandemic. thecamp receives the largest part of revenues from large companies. MakeSense remains an exception with its transparency. The 2020 annual report is available on its website: it reports that 6% of the total €5.3 million global (consolidated) revenue comes from foundations and philanthropy, 31% from public subsidies, 19% from corporate sponsoring, and 44% from services to large companies.

The dynamics of the evolution of open labs shows that complementarities in revenue sources have now become the norm everywhere. Open labs follow the money with pragmatism and, eventually, forget that the coworking space was initially intended to host innovation, creativity, or incubation projects. This necessary pragmatism sometimes generates concerns about the strategic alignment between the portfolio of services, the list of paying customers, the associated pricing schemes, and the strategic intent claimed by the open lab.

2.2.3 Operation model: how to cover fixed costs?

The operation model describes humane, physical, and technological capabilities, and the associated processes. It is easy to describe the list of capabilities required to run strategies in the "thinker" versus "maker" orientations. The description of the service model already provides the implicit elements to understand operations behind the services.

Open labs operating in the "thinker" orientation must install competencies providing innovation- and creativity-related intellectual services (including incubation and acceleration), and the associated facilities. The analysis of facilities elaborates on costs incurred by the space used by start-ups, coworkers, and staff. It is most often a flexible open space that can be reconfigured to host events. The operational model therefore runs on two legs: fixed costs for facilities, fixed and proportional costs for human resources and machines.

In the domain of human resources, it is always possible for open labs to rely on external competencies for a part of their services, but they need to host a permanent staff suited to offer innovation and creativity-related services, and to organise the associated events. Open labs can easily set up a network of experts, mentors, coaches, and event facilitators who can only show up when the open lab needs them. When operating in this way, open labs must cover the fixed costs of permanent staff with recurrent revenues, while they tailor the use of external staff to the occurrence of events (and of associated revenues). Open labs can therefore operate with a significant part of variable costs proportional to the volume of start-ups, or of events. As illustrated by Liberté Living Lab, open labs can also reach out to experts in devoting specific workstations in their open space in exchange for services delivered to their community. Open labs investigated during field research never encountered any difficulty in relation to the staffing of such competencies. All of them managed to easily ramp up the size of their network of experts, coaches, and mentors to meet the scale of their activities. The staffing of activities in open labs was also easy because they have used all possible subsidised jobs available in France to support their activities, most notably with interns, work-study training grants, and "service civique". The latter is a voluntary service established in 2010 for young people between the ages of 16 and 25 after the end of conscription in France. Aiming at reinforcing national cohesion and social diversity, it is funded with a state-provided monthly allowance. Open labs are eligible for all three modalities and can therefore expand their staff at limited cost.

The management of human resources and facilities applies to all open labs, but the specific concerns incurred by the presence of machines only apply to open labs in the "maker" specialisation. This translates into specific constraints on capital in the business model. "Up-to-date" services in relation to prototyping, or with the production of a short series of products, require a large sample of technology-intensive machines. High service quality and cost effectiveness also require working with up-to-date technology. Several open labs, such as La Paillasse, often use second-hand machines purchased from, or gifted by, industry or research agencies. Make ICI and ICI Montreuil buy new or nearly new machines to propose attractive services to their specific clients, start-ups, and SMEs. Les Garages XYZ in Saclay, with a specialisation in pre-industrialisation, face even more stringent constraints. This type of business positioning remains fragile and complex: reaching a positive EBITDA or breaking-even proved to be almost impossible in Usine IO's case. The required competences expand to the management of safety and security issues, including training for users and incubated start-ups. In these open labs, the business model has therefore to address specific profitability issues and threshold effects related to the use of each machine, and to the portfolio of machines. Business models must therefore consider all standard issues faced by any other company relating to acquisition costs, amortisation, depreciation, replacement of machines, and maintenance (Figure 2.2).

Zooming out from the operation model, it is possible to identify a list of critical resources relevant for all open labs: facilities, technology and machines, and human resources. Fixed costs prevail in the business models. Consequently, open labs confront three strategic issues:

- **Saturation of occupancy rates:** occupancy rates represent the main driver to run an open lab business. It applies to workstations in the

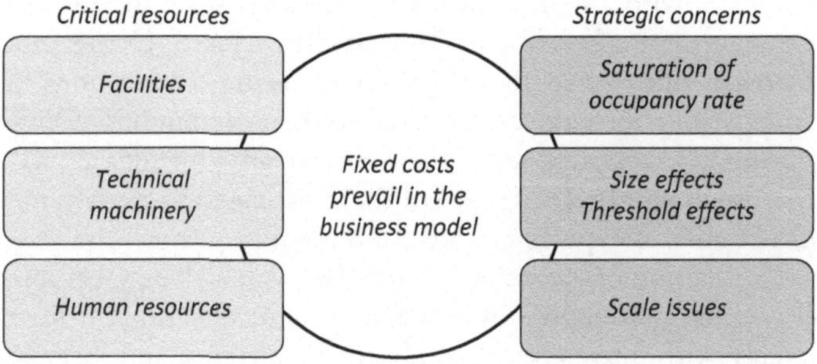

Figure 2.2 Critical resources and strategic issues in the business model.

coworking space, and to the use of machines. Improving occupancy rates therefore requires an explicit ability to find opportunities for complementary services and revenues with related clients in a related diversification strategy that neither distorts the mission of the open lab, nor its legitimacy with innovation or creativity services, nor its attractivity to target clients who are parts of its community. In a nutshell, related diversification cannot crowd out target clients or endanger strong ties with the community because of the need for cash, or it will kill the strategy of the open lab.

- **Threshold effects:** a difficult point is to seize the capabilities to meet the appropriate level of effective demand (demand that generates effects and translates into revenues) in medium- and long-term timeframes. As in any other firm, open labs need to adapt their capabilities to the level of demand consistent with the recurrent flux of activities in the ecosystem. Excess capabilities will automatically translate into irrelevant fixed costs and threshold effects, thus endangering the permanence of each open lab in its ecosystem if capabilities are seized to meet short-term artificially high levels of demand, or if they poorly anticipate the evolution of competition between open labs. Threshold effects directly depend on size, and on the number of open labs in the ecosystem. The typical threshold effect can be illustrated in the Paris innovation ecosystem: when Station F had installed its activities, its cost structure made it possible to offer workstations in its incubators for a very low price that disrupted competition, thus automatically killing one of the few options used by the other open labs to survive in the short run while adapting their service models.
- **Scale**: the scale issue emerges when the coverage of fixed costs requires scaling up the portfolio of services and the volume of clients. It has been typically met by open labs that were initially thinking about relying on the sale of stakes in successful start-ups to pay for regular operation costs. NUMA's story before its last transformation proved the difficulty with such a strategy or its eventual inconsistency in the French (business and financial) environment because of the gap between inflows and outflows. The scale issue materialises with two different questions. First, what is the appropriate level of capital for the open lab operating this strategy? Second, what is the size of working capital required to feed the ramp-up of activities towards the moment where inflows match outflows? Considering the normal attrition rate for the development of start-ups in each sector, and the difficulties in articulating basic research with the development of start-ups in the French innovation environment despite all public policies and the proactive actions carried out by Bpifrance, the scale issue meets a problem linked to the size of ecosystems or to the volume of candidate start-ups that can only be addressed with geographical diversification or with thematic specialisation. The scale issue then transforms into a scalability issue.

At this stage of the analysis of open labs in the French (or European) environment, it is easy to identify that the profitability of an open lab's business model depends on the interplay between size, threshold effects, and occupancy rates. This means that it follows rationales traditionally prevailing in the "old economy". However, there are lots of different ways to cook a business model for an open lab, and no magic formula to achieve sustainability or profitability is available.

The possibility of breaking even remains an open question for most independent open labs unless these initiatives are inserted into broader considerations. Life is obviously easier when these initiatives are carried out by public or private actors who have other perspectives in mind. Most open labs in the sample, like MakeSense or the Make ICI network, started on a small scale before ramping up progressively. Other open labs directly tried to cope on an intermediate scale, like NUMA or Liberté Living Lab. Other labs in the sample have directly kicked off their activities on a very large scale. They are coined as "mega open labs" or "mega innovation platforms" in Chapter 1 and will be discussed now.

2.2.4 Specificities of mega open labs: kicking off on a very large scale

Very large open labs are not numerous in Europe and in France. Field research has identified several cases atypical by their size in France (Mérindol and Versailles, 2017): Station F in Paris, thecamp in Aix-en-Provence, and Bel Air Camp in Lyon. These cases illustrate the specificities of size effects and push the reasoning about the coverage of fixed costs to limits that prove to be incommensurate with the other open labs investigated in this chapter.

Station F and thecamp illustrate a specific story and an original evolution, with unique entrepreneurial ventures promoted by "missionary" entrepreneurs (Gruber and McMillan, 2017). Bel Air Camp has an original story in Villeurbanne, in the Lyon suburbs. Only NUMA or Liberté Living Lab which operate their own five or six-floor buildings in the Sentier district, in the centre of Paris, could somehow compare to such sizes. Most standard-size open labs investigated in this chapter (e.g., MakeSense, Liberté Living Lab) incubate c.15 start-ups per intake, with a maximum of two intakes per year. Station F remains an emblematic instance of the size effect: its communication aggressively promotes the figure of 1,000 start-ups under incubation. When other open labs in the sample offer 50 to 150 workstations, in a 1,000 m^2 coworking space, Station F offers more than 3,000 workstations over 34,000 m^2. In comparison, the public mega open lab Euratechnologies operates an 80,000 m^2 facility in Lille. Bel Air Camp currently uses only the ground and first floor of a former factory; the other floors are available for future expansion. Mega open labs do not compare to other open lab instances investigated in this chapter for the size of facilities, the volume of workstations, and the total amount of capital expenditure and of working capital required to run these projects.

Initially established in September 2017, thecamp results from the personal initiative of Frederic Chevalier, a successful businessman who passed away in an accident some weeks before kicking off the open lab's activities. The project depends on a part of his personal fortune made available through a holding company and a foundation. Before passing away, Chevalier had time to buy the land and build the facilities through his holding company: several futuristic buildings for operations on the Arbois plateau close to Aix-en-Provence TGV station, and 160 hotel rooms to host artists-in-residence and visitors. Chevalier also injected enough money into thecamp to cover the working capital. thecamp has recorded a total revenue in excess of €6 million for 2021, a "bad" year because of the Covid pandemic, compared to €10 million in 2018 and 2019. thecamp never covered its costs before the beginning of the Covid pandemic and was searching for possible reconfigurations in its service model and its pricing schemes, thus reshaping interactions with the large firms that also represent its strategic partners and clients. The lockdown and the pandemic made it necessary to update the service model and recapitalise the company. A new roadmap was recently made public in the press by Olivier Mathiot, the newly established non-executive president (*La Provence*, October 5, 2021). The holding recapitalised the project, and negotiated loans from recurring partners, most notably Crédit Agricole bank that is already running one of his incubators, Village by CA, in thecamp. The staff operating the open lab had around 50 FTE before the pandemic; it is now slightly lower after several staff members left the project. Mathiot discloses that the total of recapitalisation plus the reduction of operation expenses through the renegotiation of rents and loans amounts to €20 million. The optimisation of OPEX was made possible because land and facilities are owned by a foundation that is itself affiliated with Chevalier's holding, which has a vested interest in supporting thecamp's mission. The evolution of activities went in several directions: put several programmes on hold until profitability resumes, and travel restrictions incurred by the pandemic are lifted; refocus on the short run on profitable programmes for large companies; organise a new division of activities with the incubator Village by CA and with the makerspace operated by Make ICI; and split hospitality activities from the open lab with externalisation to a new subsidiary in charge of operating it as a standard hotel open to any client, not only to incubated start-ups, artists-in-residence, or employees from large companies visiting for research projects. Mathiot explains that this reorganisation makes it possible to "divide the former break-even point by a factor two".

Station F in Paris relates to other rationales. During the two initial years of its activities, Station F received more than 11,000 applications from start-ups for all its programmes; the selection rate was 9%. Over the same 24-month period, about €250 million was raised by 232 start-ups affiliated with them. This should be compared to the €150 million raised by 148 start-ups accelerated by NUMA over a ten-year period. Significant investments were made to launch Station F in September 2017: an article published in *Challenges* (on June 19, 2013) mentions €70 million to buy the facilities from the City of Paris, and €60 million for its refurbishment, partially supported by the French government through the Caisse des Depôts. The conversion of La Halle Freyssinet industrial wasteland articulated with local public policies promoted (and financially supported) by the City of Paris. The press mentions a total investment of €250 million by Xavier Niel, the very successful entrepreneur who owns (among others) Iliad, the parent company of the French telecom provider Free, and the international seed fund Kima Ventures. Niel is considered one of the wealthiest French people; he appears regularly in the top 250 of the world's wealthiest individuals. The only incubation programme directly managed by Station F is the Founders Programme. The economic weekly *Challenges* reported (October 10, 2018) that Station F attracts lots of criticism because it acts as a broker (Agogué et al., 2013) focused on early-stage incubation and event organisation without providing actual incubation services. They connect, for instance, with services provided by Paris Chamber of Commerce. They have actual contributions to make facilities available and to run an aggressive communication programme about the Station F brand. *Challenges* also points out the poor quality of hospitality services during the years 2017–2018 (i.e., during the ramp-up of activities).

Station F organises activities in three silos, respectively dedicated to events and networking ("Share"), to spaces dedicated to the 25+ corporate incubators hosted in Station F premises ("Create") and facilitated by staff employed by these companies, and to hospitality or restauration ("Chill"). With the exception of the Founders Programme, all incubation-related activities are operated within corporate incubators buying dedicated spaces in a B2B mode and covering costs and risks for their respective start-ups. Station F's business model follows standard B2B real estate renting at zero risk. Only start-ups hosted in the Founders Programme directly pay their fees to Station F. Corporate incubators pay a rent to Station F irrespective of the number of workstations used. Size effects also make it possible to charge low rates: €200 per workstation per month, while other open labs in Paris used to charge €350 to incubated start-ups. Station F has scaled activities at levels previously unmet in a for-profit business model. The reality emerges progressively. Companies that only purchased access to networking complain that the "Create" silo is hermetically closed and managed against the rationales of open innovation. Start-uppers complain that Station F is no longer adaptable after the early-stage phase. Companies hosting their corporate incubator in Station F are satisfied with services and with the visibility granted by Station F's brand. Other companies realise after paying their fees that Station F operates as a platform hosting incubators, not as an "open lab" itself.

Station F and thecamp illustrate the advantage of kicking off activities at the proper scale. While other open labs must fight for cash to build their ramp-up day after day, thecamp and Station F started operations from day one with cash, facilities, and working capital. This does not mean that their business model does not need adjustments. thecamp clearly had(has?) a fragile position as a mega open lab. However, the back-up of Chevalier's personal holding and foundation, now run by his heirs, made it possible to survive and adapt. Station F looks like a start-up factory, but its business is one of a B2B real estate manager: it sells indirect services required for incubation, and mostly hosts incubation performed by others. In both cases, but for different reasons, the business models articulate with much broader considerations located in holding companies owning their facilities and their brands. It is therefore impossible to appraise the business model of Station F and of thecamp without zooming out and considering the different advantages generated by these activities in their respective holdings. Such an optimisation scheme might apply to smaller open labs, or smaller ventures, but the size effect available at the kick-off of activities definitively provides a competitive advantage.

2.3 From scale to sustainability

The previous developments show how fragile the business models are in open labs. Open labs have it difficult collecting cash and stable revenues from their targeted clients, and at working with the appropriate level of working capital. This section discusses root mechanisms projecting the dynamics of tangible and intangible resources into future developments.

2.3.1 Scale, fixed costs, and working capital

The different components of business models show that the management of tangible resources and of human resources in open labs follows very standard rationales: the search for break-even points depends on threshold effects linked to fixed costs (facilities and employees on the payroll). The saturation of occupancy rates represents the most visible managerial target, but the investigation of threshold effects reveals that the tension between scale and sustainability is the most important issue. There is no "good" response when trying to identify the appropriate scale to launch or operate an open lab project: as in any other company, successful open labs show consistency between capital and working capital, the volume of activities, and the ability to cover fixed costs. The management of fixed costs and the availability of appropriate levels of working capital unsurprisingly explain sustainability issues.

Projections about the ramp-up of activities and potential developments should also consider effective demand and sensitive points about the boundaries of the client base, thus leading back to the discussion of the appropriate scale to be used for an open lab project. Make ICI and MakeSense, and Station F and thecamp even more, illustrate business models where the client base has both local and global roots. Make It Marseille, conversely, only anchors in local ecosystems and opted for smaller scales. Strategic diversification options and possibilities to adapt to threshold effects are more limited in the series of "smaller" open labs, precisely because of their local boundaries. The scale challenge is framed by the respective sizes or boundaries of the client bases. An important point is also that the number of open labs in each ecosystem cannot grow *ad libitum*. All these ventures compete for the same clients and offer similar services unless they specialise in topics, or phases of the incubation process, or services (e.g., makers versus living lab-type experimentation). This means that discussion of the open lab's scale also depends on the global size of its reference ecosystem.

Field research reveals an insidious difficulty with the management of the service portfolio. Open labs have adapted to meet clients able to afford their services: they have followed the money. The business model therefore encounters an explicit difficulty with diversification strategies. It makes sense to expand service portfolios towards related services and related clients unless the alignment between strategic intents and sources of revenue is at stake. Economics of scope and diversification are easy to compute in the open labs operating "thinker"-related services. The challenge is different for open labs operating in the "maker" orientation. Expanding towards related clients with the same machines, or with additional "thinker"-related services, is not difficult in their case. The acquisition of additional machines is conversely an uneasy strategic move, thus limiting the perspective for diversification.

Beyond this almost trivial evidence, major issues arise when strategic intents and sources of revenue do not align anymore. It is most specifically the case for open labs that allegedly target services for non-profit or social-business communities, and insidiously follow the money available from large companies. Such incompatibility issues may put the whole open lab project at stake as they endanger the dynamics of the community. This point ultimately shows that the management of fixed costs does not exhaust the analysis, even if fixed costs represent the main important reference to manage daily activities in an open lab.

2.3.2 Strong ties, communities, threshold effects (again), and scale

The second aspect of the discussion focuses on intangible resources and competencies. The discussion of human resources has already identified that a part of the competencies required to run open labs can be hired on a case-by-case basis, and therefore represent a cost commensurate to activities where the open lab needs experts, coaches, mentors, event organisers, or facilitators. This part of the staff will hopefully build close relationships with the rest of the open lab, and with permanent staff members. Provided that enough permanent resources are devoted to the organisation and facilitation of such a network of contributors, it remains easy to anticipate that these people will develop strong ties with the open lab's staff and will adhere to the same values or ways of working. In this respect, the evolution of costs in the open lab therefore follows the volume of activities. If the size of the ecosystem permits, it is even possible to anticipate that a significant growth in the volume of activities will attract more talents.

All open labs, whatever their sizes, struggle to cover their costs but they do not have issues in finding clients or at populating their communities of users. They only have concerns with clients able to afford their services because the management of innovation and the development of entrepreneurial ventures makes it difficult to pay for services proposed by open labs. Large companies are most often the only category of clients able to pay for the open labs' services. However, zooming out from the management of scale and fixed costs, it turns out that the most important driver of the sustainability of open labs resides in the dynamics of their community. Two points explain this relevance: interactions with the management of open labs, and interactions among community members. Both aspects build an open lab's brand, which is embodied in its facilities (these in turn become some sort of totem place, see Mérindol and Versailles, 2017).

The analysis of interactions between the open labs staff and community members demands flexibility and availability. It also materialises in the ability to adapt the network of experts, coaches, and mentors when needs evolve inside the community. The proximity between community members and experts, coaches, mentors, or staff members represents a key issue. From a managerial point of view, this translates into a managerial key indicator: the number of projects followed by each expert, mentor, or coach. From a conceptual point of view, geographies of knowledge make it possible to understand the role of cognitive proximity and of the vivacity of cognitive links in the business model of open labs. The nature of interactions is characterised by weak versus strong ties (Granovetter, 1973). Interactions between community members and the nature of ties obviously differ with the type of communities described in Chapter 1, and already used in Table 2.1. As identified by Capdevila (2017) in his research on makerspaces and on coworking spaces focused on social orientation, strong ties develop with the combination of four complementary mechanisms: local and global information and communication ecology (Bathelt et al., 2004), and local and global pipelines that articulate with frequent and continuous knowledge flows mixing face-to-face and virtual interactions.

Co-locating individuals sharing the same practices, or similar activities and projects, in the same facilities, does not ensure success. Communities depend on a high degree of proactivity and openness in knowledge-sharing activities. Further research will be required to understand the conditions of success for communities. However, the evolution of open labs investigated here shows that their business models are orchestrated around communities. When communities of open labs lose their strong ties and transform beyond weak ties, the open lab institutionalises; the service model then only focuses on innovation-related consulting. This is illustrated by NUMA's progressive transformation. The elaboration of the Make ICI network conversely illustrates how the business model of an open lab can articulate local and global pipelines, and local and global communities (or "buzz" in Batheld et al.'s (2004) words). When loose cognitive ties emerge from the dynamics of communities, open labs need some form of rejuvenation, creating value for their users. This chapter therefore shows that the business model of open labs directly depends on the vivacity of their communities. Their respective life cycles are totally intertwined.

The analysis of costs has identified the importance of threshold effects. It is now possible to interpret them under the perspective of community lifecycles. When confronting new experts, mentors, and coaches with different competencies, staff members in the open labs will improve their ability to support projects and provide community members with solutions. When increasing the size of the network of experts, coaches, and mentors, the ability to address new problems will improve. All these aspects elaborate on the local and global buzz, and on pipelines. Virtuous circles will ensue, and the impact of the open lab's brand will follow. There is no direct limit to the size of the network of coaches, experts, mentors, and to the volume of staff members in an open lab. However, there is an obvious limit to the nature of links inside a community, as there are limits to the "buzz". Boisot (1998) has already explained that "social learning cycles" develop from a mix of tacit, codified, and abstract knowledge, thus framing communication and the diffusion of information (or knowledge). The economies of knowledge and information transmission (as in Boisot and Li, 2006: 129–130) frame the dynamics of communities, and therefore the lifecycle of their business model.

It therefore becomes difficult to assess an open lab's critical mass, because the characterisation of the critical mass of a community is very difficult to assess. From a business model perspective, positive externalities and virtuous effects will emerge from the growth of the network of coaches, mentors, and experts. The same applies to the potential portfolio of competencies available in the open lab if it can cover the associated costs. The visibility of the open lab and its brand will then grow as well. However, the sky is not the limit. The eventual parallel growths of the community, of the network of experts, coaches, mentors, and of the volume of staff, will soon confront the moment where cognitive links vanish. To use Granovetter's words, strong ties will first transform into weak ties and then progressively vanish. The business model of open labs then entirely depends on the ability to cook a ramp-up towards a scale that makes it possible to cover fixed costs, while preserving strong ties inside the community. As illustrated with Figure 2.3, the analysis of strategies and the potential emergence of economies of scale and/or of scope needs to cope with the tension between the volume of tangible and human resources (leading to an analysis of costs and break-even points), and intangible resources.

2.4 What's next?

Open labs have proven the originality of their contributions for the management of creativity and innovation, but their business models remain fragile.

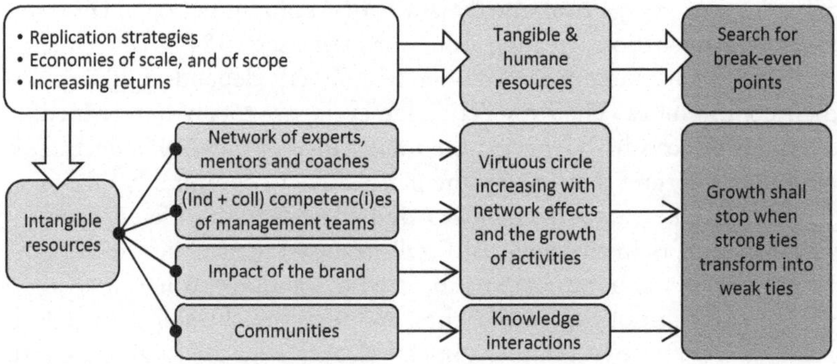

Figure 2.3 Tension between tangible, human, intangible resources, and communities.

The sustainability of their business models cannot be managed without paying close attention to scale issues and working capital. However, cost-benefit analysis does not exhaust the reasoning because the open lab's attractivity depends on intangible aspects rooted in the dynamics of its community, and on the alignment between the open lab's strategic intent and its community.

Field research and more than a decade of evolution demonstrated that services delivered by open labs have a distinctive value in ecosystems. Open labs therefore have a future. However, the added value of open labs does not automatically translate into solutions to break even and cover the costs of tangible and human resources because scaling up activities has limits. The sustainability of the business model depends on the dynamics of knowledge in the community (social learning cycles, local and global pipelines, local and global "buzz"), and therefore on the survival of communities. It is totally impossible to separate their respective trajectories.

Is it difficult for an open lab to nurture a community? Obviously not. Open labs listed in this chapter all instantiate effective services delivered to communities. Where does the critical mass for a community in open labs lie? Only considerations about social learning cycles and the dynamics of communities can tell. How can open labs survive without a community? It will be difficult for an open lab to run a sustainable business model if it is not able to cover its costs when the community is alive, but it will be impossible to keep the open lab alive without community members feeling that they benefit from an active community life, and from the quality of knowledge exchanges inside the community. Is there a life for open labs after the death of their community? Probably not. Open labs will find it difficult to create value and differentiation beyond genuine consulting if their community dies.

Bibliography

Agogué, M., Yström, A. and Le Masson, P. (2013). "Rethinking the role of intermediaries as an architect of collective exploration and creation of knowledge in open innovation". *International Journal of Innovation Management*, 17(2), pp. 1–24.

Andreini, D., Bettinelli, C., Foss, N. J. and Mismetti, M. (2021). "Business model innovation: A review of process-based literature". *Journal of Management and Governance*. Retrieved August 14, 2021, from 10.1007/s10997-021-09590-w.

Barney, J. (1991). "Firm resources and sustained competitive advantage". *Journal of Management*, 17(1), pp. 99–120.

Barney, J. B. and Hesterly, W. (2005). *Strategic management and competitive advantage*. Upper Saddle River, NJ: Prentice Hall.

Batheld, H., Malmberg, A. and Maskell, P. (2004). "Clusters and knowledge: Local buzz, global pipelines, and the process of knowledge creation". *Progress in Human Geography*, 28(1), pp. 31–56.

Boisot, M. H. (1998). *Knowledge assets: Securing competitive advantage in the information economy*. New York: Oxford University Press.

Boisot, M. H. and Li, Y. (2006). "Organizational versus market knowledge: From concrete embodiment to abstract representation". *Journal of Bioeconomics*, 8(3), pp. 219–251.

Budler, M., Zupic, I. and Trkman, P. (2021). "The development of business model research: A bibliometric review". *Journal of Business Research*, 1935(June), pp. 480–495.

Carroll, A. B. (1991). "The pyramid of corporate social responsibility: Toward the moral management of organisational stakeholders". *Business Horizons*, 34(4), pp. 39–47.

Capdevila, I. (2017). "The local and global knowledge dynamcis through communities: The case of communities of makers and social entrepreneurs in Barcelona". *Management International*, 21(3), pp. 59–70.

Elkington, J. and Hartigan, P. (2008). *The power of unreasonable people: How social entrepreneurs create markets that change the world*. Boston, MA: Harvard Business Press.

Foss, N. J. and Klein, P. G. (2012). *Organizing entrepreneurial judgement: A new approach to the firm*. Cambridge: Cambridge University Press.

Foss, N. J. and Saebi, T. (2017). "Fifteen years of research on business model innovation: How far have we come, and where should we go?" *Journal of Management*, 43(1), pp. 200–227.

Foss, N. J. and Saebi, T. (2018). "Business models and business model innovation: Between wicked and paradigmatic problems". *Long Range Planning*, 51(1), pp. 9–21.

French Law #2014-856 published on July 31st, 2014.

Granovetter, M. S. (1973). "The strength of weak ties". *American Journal of Sociology*, 78(6), p. 136080.

Gruber, M. and McMillan, I. (2017). "Entrepreneurial behavior: A reconceptualization and extension based on identity theory". *Strategic Entrepreneurship Journal*, 11(3), pp. 271–286.

Johnson, M. W., Christensen, C. M. and Kagermann, H. (2008). "Reinventing your business model". *Harvard Business Review*, 86(12), pp. 15–36.

McMullen, J. S. and Bergman, B. J. Jr (2017). "Social entrepreneurship and the development paradox of prosocial motivation: A cautionary tale". *Strategic Entrepreneurship Journal*, 11(3), pp. 243–270.

Merindol, V. and Versailles, D. W. (2017). *Créer et Innover aujourd'hui en Ile de France : le rôle des plateformes d'innovation*. Research co-funded by Bpifrance Le Lab et Innovation Factory, newPIC Chair, Paris School of Business. http://www.newpic.fr/02proj2016openlabsidf.html.

Mérindol, V., Versailles, D. W. and Le Chaffotec, A. (2022). "Les organisations intermédiaires et l'innovation : les multiples facettes de l'intermédiation de réseau". *Innovations*, 2(65), pp. 49–80.

Santos, F. M. (2012). "A positive theory of social entrepreneurship". *Journal of Business Ethics*, 111(3), pp. 335–351.

Short, J. C., Moss, T. W. and Lumpkin, G. (2009). "Research in social entrepreneurship: Past contributions and future opportunities". *Strategic Entrepreneurship Journal*, 3(2), pp. 161–194.

Part 2

Open labs as innovation intermediaries

3 Art, entrepreneurs, and open labs

New challenges to foster open innovation

Nicolas Aubouin

Open labs (Schmidt and Brinks, 2017; Mérindol et al., 2016) such as fab labs, living labs, hackerspaces, and makerspaces have multiplied over the recent years. Specific open labs which can be termed as "open art labs" have also emerged in France (Aubouin, 2018). They can be defined as spaces of artistic and cultural creativity, in the sense of the collective production of new and original ideas recognised as useful and relevant to organisations (Anderson et al., 2014; Amabile et al., 1996). This type of open lab develops in the interstices between organisations (Cohendet and Simon, 2016), either within *ad hoc* spaces in companies or cultural institutions, or in autonomous spaces outside these organisations. They question the boundaries of organisations by developing multiple and varied collaborations between different stakeholders (Aubouin and Capdevila, 2019). In Aubouin (2018), they are defined as "spaces and approaches that question the role of artists, entrepreneurs and engineers around the renewal of creation, production and dissemination modalities; collaborative, iterative and open processes; multiform communities and the mobilisation of different user profiles".

From a wider perspective, open art labs highlight the articulation between the artistic, technological, and entrepreneurial dimensions within open innovation spaces. These physical spaces and services can take various forms and reveal specific dimensions regarding the roles that artists can play in open innovation. In fact, open art labs can be first of all analysed as trading zones (Galison, 1999; Wilson and Herndl, 2007) i.e., places for exchange and negotiation between artists, engineers, and entrepreneurs (Hargadon and Bechky, 2006) around a common creative project. The physical space then conditions the capacity to generate new ideas (Dougherty, 2001; Vyas et al., 2009), to confront them, and even to exploit them in a relevant way to develop the project. This physical space can then appear as an interstitial space (Furnari, 2014) i.e., as a temporary space that allows for the generation of shared meanings and new practices that will spread beyond the boundaries of this space. These spaces facilitate the emergence of new practices. They become catalyst places: they allow the coordinated creation of new practices, or new knowledge, from the diversity of the members who meet there. The space can also be analysed as a resource in its own right (Eisenhardt and Schoonhoven 1996): it is not only a learning space,

DOI: 10.4324/9781003125587-6

articulating key knowledge assets and skills for the project, but also constitutes a key resource mobilised by the community that develops there.

From this point of view, the open art lab is intimately linked to the community of innovation (Peredo and Chrisman, 2006; Amin and Cohendet, 2004) which integrates and deploys various projects. To understand the functioning of the community, it is necessary to analyse the role of its members and, in particular, in the case of open art labs, the specific contribution made by artists to entrepreneurial projects with their different characteristics (Figure 3.1).

This chapter will investigate this issue: it will analyse the role of artists as innovators (Sternberg, 1999; Oakley, 2009) in open labs. How are the artistic, technological, and entrepreneurial dimensions articulated within these open labs? What are the specific contributions by artists to open innovation processes that emerge and unfold in open art labs?

To answer these questions, we develop a case study based on three French open art labs: the 104 Factory, Artlab, and Atelier Arts Science. This comparison highlights the issues and the trajectories of the organisational construction of open art labs around the artistic, entrepreneurial, and technological dimensions as well as the diverse contributions by artists to open innovation processes. After having introduced three types of possible roles for the artist in an entrepreneurial project, we show how these roles can be integrated into open labs with reference to three illustrative case studies, and then investigate how to think about the coherence between the open lab, the entrepreneurial project, and the role of artists.

3.1 The artist as innovator in open art labs: three roles to articulate art, technology, and entrepreneurship

Creative spaces are intimately linked to an innovation community that fosters the development of creative practices (Peredo and Chrisman, 2006; Amin and Cohendet, 2004). To understand how a community behaves, it is necessary to analyse not only the role of community members but especially, in the case

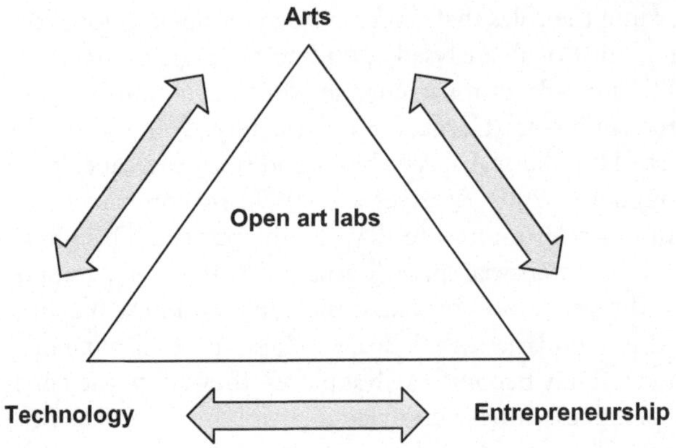

Figure 3.1 Open art lab as a space for articulating arts, technology, and entrepreneurship.

of an open (art) lab, the role of artists as key players in the community for articulating the artistic, technological, and entrepreneurial dimensions.

The artist present in open labs will therefore be analysed as an actor of creativity and innovation (Sternberg, 1999; Oakley, 2009), and as an entrepreneur of the project (Menger, 2002). Brechet et al. (2009) provide a typology of the three aspects of the entrepreneur and innovator: visionary, expert, and relation-builder. We use this typology to develop a dynamic approach of the role of artists as entrepreneurs and innovators. They evolve with the artistic project and the organisational configuration in which it unfolds.

The role of artists as innovators (Sternberg, 1999; Oakley, 2009) is to promote a new vision, to create bridges between different worlds, therefore acting as boundary spanners (Levina and Vaast, 2005) and articulating artistic, social, and economic expertise. They behave as co-producers of innovation (Imbert and Chauvet, 2013).

3.1.1 The artist as a visionary: from innovator to boundary spanner

Scholars in sociology and art management investigate the artist's role as one of an entrepreneur who promotes artistic projects and develops specific capacities for risk-taking and innovation (Menger, 2002).

From this point of view, the artist can also be considered as an entrepreneur who, through his original ideas, poses as a deviant and upsets the established order. This role casts the artist as a "romantic" figure (Chiapello, 1998) who creates an original worldview and questions established rules and norms. This innovative role has been widely studied within the creative industries (Sternberg, 1999; Oakley, 2009) but it is usually not specifically investigated in the context of public art institutions.

The role of the artist as innovator can be analysed from two different conceptualisations: the artist as a worker (Menger, 2002) or as a visionary entrepreneur (Brechet et al., 2009) who imagines and conceives his project in an original manner. The visionary entrepreneur possesses a unique capacity to discover business opportunities and to implement them. The challenge for entrepreneurs is to also ensure that their vision is understood and shared by the actors through whom their project will be formed and realised.

However, the authors cited above insist that a relational substrate is necessary for the acceptance of a visionary entrepreneur's ideas. Thus, the creative ideas of the entrepreneurial artist can be disseminated only if he can count on the support of a few well-placed connections among the networks of the economic and art worlds when faced with opposition and risk (Becker, 1988).

The relational dimension represents a central dimension of any entrepreneurial approach. It also describes a full-fledged figure: a relational entrepreneur (Brechet et al., 2009) who plays the role of a network player by enrolling actors around a project that could be described as "generic". Thus, if the entrepreneur's social and relational capacities are strongly solicited, he acts as a mediator to enable the actors engaged in a collective action to recognise each other's skills, pool their knowledge, and engage in a process of building shared rules that constitute the common project.

From the perspective of creativity management, the entrepreneurial role could be compared to a boundary spanner (Levina and Vaast, 2005), connecting individuals from different worlds, facilitating exchanges, and helping convey knowledge from one context to another (Tushman, 1977; Lingo and Mahony, 2010). In the context of artistic projects, the boundary spanner contributes to translating the artistic project into an entrepreneurial language, he builds relationships between actors with heterogeneous skills (Goglio-Primard and Crepin-Mazet, 2011; Espinosa et al., 2003).

3.1.2 The artist as an expert: from boundary spanner to co-producer

As we have seen through his double role as visionary and relational entrepreneur, the artist, by creating connections between worlds, is the driving force behind the construction of the value proposition and even the creation of a model of "innovative business". He not only produces new ideas but also new prospects for technical and even commercial development. Thus, he is the bearer of not only a vision, a connection, but also a specific expertise.

According to Brechet et al. (2009), the artist as entrepreneur is therefore an expert, the bearer of specialised, legitimate, and recognised knowledge. According to those authors, he ensures innovation by proposing genuine creative responses and by favouring the adoption of new combinations of production factors. However, by mobilising the work of Mustar (1994) on entrepreneurial researchers, the authors specify that the expert entrepreneur is not an isolated creator, demiurge, or "heroic entrepreneur" who, with the possession of his knowledge, can realise his project in isolation. Instead, the success of his approach is based on a process of entrepreneurial socialisation, implementing real interfaces of relational mediation which offer mechanisms for the networking of pluralistic actors, sometimes as carriers and vectors of relationships and knowledge complementary to that of the researcher.

The artist is considered as a co-producer, who has managed both the creative project and the enterprise project, which is a key role in the case of open innovation. More precisely, as a node of the community, the artist plays the role of broker: that is, he creates connections between actors who usually find it difficult to meet and collaborate because of their different cultures and views of the world. Therefore, he contributes to creating a porosity between ecosystems that are composed of actors with well-defined boundaries, and generally characterised by different functions and practices.

The role of the artist in an entrepreneurial approach can therefore be analysed as a construct that largely depends on the context in which the entrepreneurial project is born and developed. Among the elements of contextual analysis, it is of foremost importance to consider the initial configuration of the lab's organisational design (Mérindol et al., 2016). We will analyse the influence of these configurations using three illustrative cases (Yin, 2009).

3.2 A qualitative study of three open art lab cases

Our analysis of the role of the artist in the open art labs is based on a study of three cases: the *Atelier Arts Sciences*, a workshop within Grenoble's CEA Tech Ideas Laboratories, *104 Factory* within the Parisian cultural institution Le CentQuatre Paris, and the *Artlab* installed by Digitalarti, a digital-art company in Paris.

These cases illustrate the different dynamics present in open art labs. 104 Factory is an example of integrating innovation laboratories or incubators for creative start-ups into major cultural institutions, such as the Louvre Museum Lab, Carrefour Numérique at the Cité des Sciences, and Erasme Museomix's initiatives in Lyon's museums. Atelier Arts Sciences is an example of establishing open innovation laboratories dedicated to art and culture in institutions or enterprises more broadly oriented toward industry and services, such as Orange Art Factory, Google Cultural Institute or the French institution called Le Cargo in Paris&Co, the development and innovation agency of Paris City. Artlab is an example of the emergence and autonomy of spaces for creativity in private companies or associations oriented toward the art market and the world of media, such as FAR/Creatis supported by the SOS group; or artistic projects in autonomous places not dedicated to art, such as the numerous initiatives carried out at NUMA, Liberté Living Lab, ICI Montreuil, and La Paillasse (discussed in other chapters of this book).

This investigation of illustrative case studies elaborates on a qualitative method with a triple approach: 15 semi-structured interviews with space managers and their users (start-up creators, partner companies), non-participatory direct observation, and investigation of secondary data (websites, documentation about spaces, etc.).

These three cases are drawn from two broader research projects covering a larger sample of French innovation open labs. Out of the 55 open labs investigated in these research projects, ten developed more specifically artistic and cultural projects. They constitute the initial sample for our analysis. We selected 104 Factory, Atelier Arts Sciences, and Artlab among these ten spaces situated in the cultural and artistic field because they matched the three roles identified in the previous section but at the same time displayed a variety of ways of working and different contexts for the emergence of entrepreneurial projects. They were selected to elaborate our taxonomy through their general coincidence and special features (Baker and Gil, 2008).

3.3 Art, technology, and entrepreneurship: three roles for artists

3.3.1 The artist as a boundary spanner: 104 Factory

3.3.1.1 104 Factory as a space articulating art, entrepreneurship, and digital innovation

In 2008, the 104 Factory was created by a cultural and artistic institution (called 104) in Paris. At the outset, the 104 Factory became an incubator dedicated to digital innovation start-ups in the field of the artistic, cultural, and creative industries. It is defined by its managing director in charge of development and innovation as an *in situ* experimentation site in interaction with the public, within the ecosystem of a multidisciplinary artistic factory, open to multiple users including artists, the public, and companies.

> When an artist comes to the 104 Factory to present his work in front of an audience, the other entrepreneurs involved in the incubator are also placed in the same situation when they present their ideas or gadgets to the public elsewhere. We support these sessions. We have experimental sessions, either individually or in teams. Sometimes, we present these artists into the 104's exhibition program. For example, we have several teams that work to create new forms of musical instruments or devices that will transform the audience's experience at a concert, for example, by improving listening. These initiatives are included in the dance show organized each month in our spaces, where one finds members of the public as well as artists and entrepreneurs.
>
> (Executive Director in charge of development and innovation)

The various stakeholders are therefore considered participants in 104's artistic programme. 104's deputy director compares the organisation to a "living lab" encouraging direct contact between artists, entrepreneurs, digital engineers, and the public through the ergonomics of spaces and the specificities of their management. In this context, the artist as artistic entrepreneur and innovator plays the role of one who transcends physical and symbolic borders, supported by the open lab's team that is in charge of several tasks. This team is obviously facilitating incubation, but it also ensures the quality of interactions within the open lab, and enforces the ability for users (spectators, visitors, residents, amateur practitioners, and passing professionals) to make remarks considered in the innovation process.

> We propose testing their devices in actual situations. For example, every month, incubated teams are invited at 104 to present their technology, give concerts to an actual audience, and test their applications and other outcomes of their projects in front of a live audience. We provide support, we work with them upstream to ensure the quality of the interaction so that everyone finds something relevant in it. In other words, we are not merely conducting marketing beta testing.
>
> (Executive Director in charge of development and innovation)

The 104 Factory is thus designed both to foster horizontal relations between its users (artists, entrepreneurs, and the public), and to erase the physical and symbolic borders around open and flexible spaces by fostering porosity among diverse artistic universes: between professionals and amateurs, between artistic and entrepreneurial projects, and between 104 Factory as an institution and its neighbourhood.

3.3.1.2 Artists as boundary spanners overcoming social frontiers between art worlds, entrepreneurship, and innovation

Artists have different contributions as boundary spanners, at various levels. They are involved in multidisciplinary projects in a physical space that promotes porosity between artistic fields with activities such as live performances, visual arts, design, etc.

> The whole project promoted by the 104 is to create value through porosity and exchange inside our facilities, and we always start with the predominant importance of artistic issues. In our facilities, we have rehearsal places, we have theatres, we have workspaces. […] And we ensure that our community members come to create and practice or to witness dance, theatre, concerts, or exhibitions. They meet thanks to artistic events, or to artistic projects. We bring them [104 users] to see other things, and we organise activities to disturb them and make them move a little bit out of their mental and physical paths.
>
> (Executive Director in charge of development and innovation)

The artist, the entrepreneur, and the public can be considered as contributors to artistic programming, but the artist-entrepreneur plays a central role allowing passage across the borders between the programmed projects:

> [*Our goal is*] to ensure that [*104*] is a place that allows for the emergence of new forms of projects stimulating creation and creativity. This applies even to internationally known artists. Our activities entangle tasks focused on the emergence, detection, and professional insertion of creators as much as artists and entrepreneurs, and we even reach out to the public to develop these tasks. […] [*They*] truly contribute to the artistic program, and this is what we promote ourselves with the entrepreneurs who come with their prototypes, their expertise. We determine how to take this rich creation potential to contribute to our programming.
>
> (Executive Director in charge of development and innovation)

Enterprising artists play the role of boundary spanners between the economic and artistic worlds through their entrepreneurial projects. This role is intimately linked to their open mindset, and their ability (or willingness) to develop hybrid projects coping with artistic and entrepreneurial constraints. The head of 104 incubator explains that these activities and these projects are supported by their role as connectors and translators. He explains further:

> What is first required is the ability to relate to these two universes [*start-ups and artists*], and purposefully locate a project within our incubator. This denotes a specific mindset. I mean, this is not obvious and shows a specific commitment. Moreover, this behavior shows that these people have an artistic and an entrepreneurial component in their brains, with a specific mindset where none of these components will ever supplant the other. Coming from different universes with different vocabularies and different objectives, [our team at 104] creates value thanks to the interaction between all these people and leaning on their respective two brains. We give equal treatment to artistic and entrepreneurial projects even while considering their specificities, and their differences as well. An entrepreneur who only meets other entrepreneurs in his everyday life may be interested in meeting people from the world of artists, but sometimes he will have a hard time crossing the bridge between these two worlds.
>
> (Managing Director in charge of development and innovation).

Therefore, the role of the open art lab (and its space) is to support the boundary-spanning role enacted by artist-entrepreneurs by organising – sometimes difficult, even improbable – encounters between artists, entrepreneurs, and the public. It is widely acknowledged that an organisation such as 104 supports or plays the role of mediator, connector, and translator, but it is not automatic that artists enact this role. This point is evidenced by the feedback collected from one of the start-up managers:

> After five months of presence in the incubator, we organised a small interaction with the public, with other artists, and with other start-ups for the economic development of our project. We brought our expertise to other projects, but the reverse was more complicated, given our slow company's ramp up.
>
> (Co-founder of a start-up incubated in 104 Factory, February 2017)

Finally, the artist as innovator and entrepreneur can transform himself within the 104 Factory in transcending the 104's physical and symbolic borders. This applies, for instance, through projects with social dimensions developed in relation to non-profit charitable organisations active in the same urban district.

> If the entrepreneur is put in touch with social and charitable organisations present in our area, he can enrich himself with the very strong creativity of people living around us. [...] These organisations and people develop great

> projects for the whole neighborhood. The 104 opens its doors and plays a catalyst role by creating bridges between all initiatives. [...] We have inverted the direction of our interaction with these organisations, by making them active contributors to 104's artistic program. We have stopped behaving like a social actor but now also act as an organisation from the world of arts and culture that builds a network
>
> (Managing Director in charge of development and innovation).

3.3.2 The artist as an explorer: Atelier Arts Sciences

3.3.2.1 An artistic approach integrated into a scientific and innovation institution

The Arts Sciences workshop originated in the early 2000s as part of CEA, the French Atomic Energy and Alternative Energies Agency (CEA) (see other comments in the Afterword). This public agency is a public research organisation with a threefold mission in scientific research, technical development, and technological transfer (classified as "EPIC" in French law). A major player in all innovation-related activities, the CEA is active in four areas: defence and security, low-carbon energy (nuclear and renewable), technological research for industry, and basic research (material and life sciences). With its recognised expertise, the CEA drives collaborative projects with numerous academic and industrial partners.

Founded in 2001, Ideas Laboratory® is an open lab hosted by the CEA in Grenoble, whose objective is to respond to future societal challenges and to drive innovation projects with high societal impact. Various disciplines are mobilised to explore, design, experiment, and mature innovations: arts, design, cognitive, human and social sciences, literature, engineering, or marketing. CEA Tech's Ideas Laboratory brings together companies, universities, local authorities, engineers, and researchers in the humanities. To develop their collective innovation processes, teams present in Ideas Laboratory mobilise artists, visual artists, live theatre artists, and designers, thus building an open art lab dedicated to open innovation with both artistic and industrial partners. They have also organised special interactions with two local cultural institutions: the *Scène Nationale Arts Sciences, Hexagone* at Meylan, and in a later stage, the *Casemate* (a centre for scientific, technical, and industrial culture).

The purpose of this open art lab and its managerial contributions seek to develop the artists' ability not only to propose a new vision of the technologies involved in their creative work, but also to confront the viewpoints of artists, scientists, and engineers to expand their creative potential thanks to technological innovation.

> *Our expertise is to organize encounters between artistic and industrial creativity domains, and* to enhance their respective potential. [...] We have a

> federative role at CEA: our job is to organize meetings [...] and to facilitate the community.
>
> (Interview with the CEO of Open Labs, CEA Tech, July 2015)

Between 2002 and 2007, several experiments occurred during a prefiguration phase where all potential contributors learned how to work together. This evolved into a partnership and the creation of Atelier Arts Sciences, uniting the worlds of art, technology, and industrial firms. In the words of the CEO of Open Labs, at CEA Tech:

> We submit a societal or technological theme to a group of artists, scientists, companies, and even journalists and researchers. We work together during their residency, which lasts from a few days to a few months. The outcome is a permanent or temporary joint project that leads to research results and often to an artistic production, a show or an exhibition.
>
> (Interview with the CEO of Open Labs, CEA Tech, July 2015)

3.3.2.2 Artists as explorers in Atelier Arts Sciences: from the capacity of shifting perspectives to innovation

The specific role of the artist in Atelier Arts Sciences can be illustrated by an emblematic case mentioned by the CEO of CEA Tech Open Labs. The meeting with Anabelle Bonnéry, the director of the dance company LANABEL, was at the origin of the creation of Atelier Arts Sciences within Ideas Laboratory.

> In 2003, Annabelle Bonnéry wanted to transform her body into a musical instrument using micro-motion sensors. She had met one of our engineers as part of [*the project*] Encounters. Within a few months of her residence, more than one hundred ideas for applications of the motion sensor technology emerged, from which we selected one dozen in various fields, which were not all applied to artistic or cultural domains.
>
> (Interview with the CEO of Open Labs, CEA Tech, July 2015)

The choreographer, through her artistic approach, initiated a show and many industrial applications. Annabelle Bonnéry contributed to innovation in two ways during this exploration phase: by pushing CEA engineers to divert from their original technological perspective and by questioning the limits of the applications of CEA's technologies. Zooming out from this specific case, her approach shows that the artist does not only question the articulation between arts and sciences while working on the emergence of creative ideas, but also contributes to the evaluation of value created during this process. Publications by Ateliers Arts Sciences explain further:

> The experience based on the body-machine relationship first revealed many questions related to the articulation of a double artistic and scientific purpose: symbolic recognition and status of the artist and scientist, research and creation. Second, this experience enabled the evaluation of the results of research carried out on the basis of a technological object diverted from its original vocation and pushed to its limits in the service of an artistic achievement.
>
> (Excerpt from the Cahiers de l'Atelier Arts Sciences, no. 1, 2007, available in French on their website – translation by the author)

The artist's approach helped to create new opportunities for economic development through certain applications of motion sensors, specifically in video games. More broadly, the role of artists in the Atelier Arts Sciences is to create what the CEA Tech's Open Labs manager calls "decentring" to describe a process leading to "shifted perspectives" and to disruption. He believes that artists in Atelier Arts Sciences put the other people present in the projects out of their comfort zones, caused a disruption against the dominant paradigm, and enabled an increase in the individual and organisational capacity to innovate.

> Innovation is no longer solely based on science and technology but also on people and uses. [...] We need even more "decentering", creativity, and discrepancy, to make us see the world differently. The artist's work is precisely to bring us his vision of the world.
>
> (Interview with the CEO of Open Labs, CEA Tech, July 2015)

3.3.3 The artist as a co-producer: Artlab

3.3.3.1 Artlab: an autonomous laboratory with artistic, technological, and entrepreneurial resources

Artlab was initiated in 2011, two years after the creation of a digital-art medium, Digitalarti, to complete its communication activities in the field of digital arts. Artlab was created as a place for production, research, development, and prototyping. As a "technological and artistic research center" (see the description available on its LinkedIn page), it hosts digital artists and "helps digital creators to make their ideas happen". The space was born out of the desire to make resources available to digital creators in the domains of technical, artistic, financial, and project management skills. This resource management activity therefore primarily addresses the challenges linked to technical and economic production of entrepreneurial projects developed by artists.

> Our Lab is […] a place for production, a prototyping place, where we think about how to develop a project, about which technical solutions to adopt, and so on
>
> (Interview with Artlab valorisation officer).

Artlab therefore represents a resource space supporting artists during the development and production of their digital artistic projects at a technical level. They benefit from Artlab's expertise in terms of technical engineering and project management, communication, and economic valorisation. The expertise provided by Artlab team members is supplemented by various technical tools available to develop the artists' digital projects: a 3D printer, a digital milling machine, a wood and metal lathe, a laser engraver, and a cutting machine. The logic of sharing human resources, machines, and networks takes the form of a business model based on service contracts with experts and timesheet-based pricing for the different tools and machines, as explained by the Artlab's valorisation officer:

> We hire people for short term missions. Sometimes we need a designer, or an electronic engineer, at another moment an engineer specialised in robotics, or a coder. We work therefore with approximately 30 people for the different projects, and we staff them ad hoc with mission-based contracts.
>
> (Interview with Artlab's valorisation officer)

3.3.3.2 The artist as co-producer: a contributor of content and network

In the Artlab case, artists play two roles. They are first co-producers of the contents of entrepreneurial projects. They also contribute to the construction/identification/elaboration of new networks of valorisation. The artistic, technological, and entrepreneurial approach within this open art lab involves the ability to renew the rationales for sharing property rights between stakeholders, thus framing a co-production model between Digitalarti and artists.

> The co-production contract acknowledges that both parties hold a portion of artistic property rights. Once the project is completed, a repartition of subsequent revenues ensues. The co-production business model is supposed to work. We really want to test it. […] We have already signed such contracts with several artists who legally own their digital art. After the contract is signed, just like with movies, the project will develop with is a director, a screenwriter, and so on. Essentially, co-production works with a revenue-sharing contract. (Interview with Artlab's valorisation officer)

In this co-production model, artists become contributors to the creative project: they are not only at the starting point of creative ideas but also act as active contributors in a large network where technicians, marketing people, production project managers, and customers are also present. These roles are shared by the other members of the Artlab, who also play a role in advising the artists, for instance about issues renewing the economic model around the artistic work, for example, by inviting an artist to abandon the classic model of art market sales and purchases to develop a rental model with partner companies.

3.4 Three modalities for open art labs: hybridisation, pollination, and pervasion

The cases described in the previous sections illustrate the artistic, technological, and entrepreneurial approaches. These three approaches highlight the complementarity issues at stake between the roles played by artists as entrepreneurs and as managers of open art labs supporting the development of entrepreneurial projects.

In 104 Factory, projects develop thanks to *hybridisation*. The framework in place for artistic and entrepreneurial projects, the way interactions and encounters are fostered in open labs and all activities related to experimentation transform the artist into a boundary spanner between the worlds of arts, the economic world of companies and start-ups, and the integration of social or societal issues when they rub shoulders with people from the direct neighbourhood. They all cross the bridges and borders between these different categories thanks to the facilitation of projects by the 104 Factory. The technological and entrepreneurial approaches are incubated by the artistic organisation while exposed to the artistic and social dimensions at the same time. In this case, open art lab managers create the conditions for encounters between these different worlds and make it possible to overcome all symbolic and physical barriers, thus allowing osmosis between ideas.

In Atelier Arts Sciences, artistic, technological, and entrepreneurial projects develop via *pollination*. Entrepreneurial projects are created when artists bring new perspectives into innovative technological projects, and make it necessary for the other project promoters, engineers, and scientists, to adopt shifted perspectives. This process of *decentring* (see Afterword) becomes the origin of a renewed entrepreneurial approach, sometimes leading to the creation of a dedicated company (or start-up). In this pollination framework, managers make an impact because they organise the connection between competencies, create meaning, and facilitate the process leading to artistic and economic value. This twofold value creation process is embodied in the parallel development of artistic projects (shows, digital works, etc.) and the creation of start-ups.

In Art Lab, *pervasion* frames the elaboration of artistic projects thanks to external technical skills, leading to co-production contracts. Pervasion also facilitates the diffusion of artistic projects toward new markets. The open art lab brings these competences to artists and promotes entrepreneurial approaches. Art Lab is thus a resource space that provides artists with an access to human and technical resources, and project management. In this case of resource allocation, the management of external relations and the construction of the open art lab's network build a sort of a "club" where "artists-as-users" are invited to work on their project during a residence. The space is then reserved for artists and engineers. It remains accessible to the public, but only upon request. This network is consequently relatively closed. Forms of cooperation in this type of open art lab are very selective, thus representing a sharp difference with other categories of collaborative spaces and of open labs that are built with an open design and with an internal logic of sharing.

The value created by the Art Lab is based on its internal expertise and on its contribution as broker of content and network (Agogue et al., 2013). This materialises with service model contracts with external suppliers, and with co-production modalities. Artistic value creation leads to economic value creation and the development of original business opportunities in different markets, namely the art market and the standard business world with companies and start-ups, all of this in interaction with public institutions. Each form of project development and of value creation (Verstraete and Fayolle, 2005) is not only carried out by artists but, also, by open art lab managers (or management teams).

In the *pollination* development case, the appropriate involvement of each category of stakeholder and the distribution of contributions to residences by artists, engineers, and entrepreneurs leads artists to adopt the innovator's posture. When leaving the open art lab's space after residences, artists become entrepreneurs of their artistic projects.

Artists can also simultaneously contribute to the development of the artistic, technological, and entrepreneurial sides of their projects. In the *hybridisation* case, the appraisal of value creation is expanded. Value now goes beyond artistic creative impacts, to encompass social impacts of entrepreneurial projects and their financial returns. Facilitation empowers the other stakeholders and creates the conditions of encounters without any other sort of active participation. Artists who limit their contributions to the facilitation of projects remain in the shadow and enter another category of contributions, most notably the pervasion modality.

In the *pervasion* case, different modalities of value creation apply to artistic projects, for instance when working with the rental of artwork, or co-production of artistic events. Open art labs will then support artists when they choose between two options: either enhance their artistic and professional recognition within the framework of an exhibition in a gallery or in a museum or promote their artwork as a communication tool rented to a company.

This taxonomy between *pollination*, *hybridisation*, and *pervasion* questions the rationales of entrepreneurial value creation (Verstraete and Fayolle, 2005). The activation of these modalities depends on stakeholders enacting creative processes in open art labs, and on the spatially situated dynamics where the artistic, technological, and entrepreneurial dimensions together articulate and feed each other. This taxonomy does not only encourage reflection about the role of stakeholders and spaces in innovative processes (Alter, 2015), but also about the economic and social impacts of creative projects.

3.5 Conclusion

This chapter sheds light on the roles enacted by artists in the innovation processes (Sternberg, 1999; Oakley, 2009): explorer at Atelier Arts Sciences; boundary spanner; promoter of a new vision creating bridges between different worlds in 104 Factory, (Levina and Vaast, 2005); and co-producer of innovation articulating artistic, social, and economic expertise in Art Lab (Imbert and Chauvet, 2013).

This dialogue between artists and open art labs allows the construction of a taxonomy of contributions by artistic projects through open art labs: 1) the parallel development of artistic and entrepreneurial projects by pollination, leading to the creation of start-ups; (2) the cross-fertilisation of mutually enriching economic and creative projects by hybridisation; and (3) the development of an artistic project pervading different methods of value creation (artwork rental, co-production, etc.). This research highlights the different issues at stake in value creation for the artistic project and for the management of open art labs, thus allowing for the identification of the subsequent roles of managers in the open art labs (compare Table 3.1).

The analysis of the roles of artists and open art labs more generally highlights the role of art to innovate, to support new entrepreneurial ventures, or to create new businesses. Artistic projects involve the creation of bridges between different worlds (Becker, 1988), reshape the boundaries between creative industries and the market of artwork, or even between arts and creative industries on the one hand and industrial and service-based activities on the other hand.

From this perspective, situating the development of entrepreneurial ventures within open art labs invites us to highlight a taxonomy with three options that remains largely questionable in the art management perspective (Chiapello,

Table 3.1 Synthesis of roles plaid by artists, open labs, and managers

Entrepreneurial and artistic dynamics	*Role of artist*	*Open labs as*	*Role of manager*
Parallel development by pollination	Explorer	Interstitial space (Cenacle)	To organise connections between expertise in creating meaning and artistic and economic value
Cross-fertilisation by hybridisation	Boundary spanner	Networks space (Agora)	To create the conditions for an encounter between different worlds
Successive development by pervasion	Co-producer	Resources space (Selectivc club)	To select and value different expertise

1998): the dichotomy between the world of art and commerce, the dichotomy between artistic and entrepreneurial projects, and the dichotomy between creative and managerial skills.

The challenge of the hybridisation of competences in open art labs also raises questions about the professional construction of a space manager and of an open art lab manager. Various studies have already provided initial thoughts on this issue, most notably about the coworking space "facilitator" (Pierre and Burret, 2014), of fab lab and open lab managers (Mérindol et al., 2016). However, the broader context of open labs and the specific role of artists in open art labs encourage further investigation of professional perspectives for entrepreneurs and engineers.

To conclude, the artist's role can only be understood as a dynamic role. The artist's entrepreneurial approach articulates around three typical options as explorer, boundary spanner, and co-producer. Contrary to the usual approach to analysing these three roles, this chapter shows that artists can successively incarnate each of those roles during the same project: the elaboration of new ideas during the project's exploration phase, and, during the exploitation phase, through his ability to translate, mediate, and connect. Artist-entrepreneurs play a transversal role with two aspects: as experts about contents, bringing specific knowledge and technical resources into project, and as a networker providing intermediation between various artistic, economic, and technical fields.

Bibliography

Agogué, M., Yström, A. and Le Masson, P. (2013). "Rethinking the role of intermediaries as an architect of collective exploration and creation of knowledge in open innovation." *International Journal of Innovation Management*, 17(2): 1350007.

Alter, N. (2015). *L'innovation ordinaire*. Paris, France: Presses Universitaires de France.

Amabile, T., Conti, R., Coon, H. and Lazenby, J. (1996). "Assessing the work environment for creativity." *The Academy of Management Journal*, 39(5): 1154–1184.

Amin, A. and Cohendet, P. (2004). *Architectures of knowledge: Firms, capabilities, and communities*. Oxford: Oxford University Press.

Anderson, N., Potocnik, K. and Zhou, J. (2014). "Innovation and creativity in organizations: A state-of-the-science review, prospective commentary, and guiding framework." *Journal of Management*, 40(5): 1297–1333.

Aubouin, N. (2018). "Dynamiques organisationnelles, modes de gestion et institutionnalisation de différents tiers-lieux culturels." *L'Observatoire*, 2: 39–42.

Aubouin, N. and Capdevila, I. (2019). "La gestion des communautés de connaissances au sein des espaces de créativité et innovation : une variété de logiques de collaboration." *Innovations*, 1(58): 105–134.

Baker, G.-P. and Gil, R. (2008). "Clinical papers in organizational economics." In: Gibbons, R., & Roberts, J. (Eds.), *The handbook of organizational economics*. Princeton, NJ and Oxford: Princeton University Press: 193–212.

Becker, H. S. (1988). *Les mondes de l'art*. Paris, France: Flammarion.

Brechet, J. P., Schieb-Bienfait, N. and Desreumaux, A. (2009). "Les figures de l'entrepreneur dans une théorie de l'action fondée sur le projet." *Revue de l'Entrepreneuriat*, 1(8): 37–53.

Capdevila, I. (2015). "Les différentes approches entrepreneuriales dans les espaces ouverts d'innovation." *Innovations*, 3: 87–105.

Chesbrough, H. (2006). *Open innovation: The new imperative for creating and profiting from technology*. Boston, MA: Harvard Business Press.

Chiapello, E. (1998). *Artistes versus Managers. Le management culturel face à la critique artiste.* Paris: Métailié Editions.

Cohendet, P. and Simon L. (2016). "Always playable: Recombining routines for creative efficiency at Ubisoft Montreal's video game studio." *Organization Science*, 27(3): 614–632.

Dougherty, D. (2001). "Reimagining the differentiation and integration of work for sustained product innovation." *Organization Science*, 12(5): 612–631.

Eisenhardt, K. M. and Schoonhoven, C. B. (1996). "Resource-based view of strategic alliance formation: Strategic and social effects in entrepreneurial firms." *Organization Science*, 7(2): 136–150.

Espinosa, J. A., Cumlmings, J. N., Wilson, J. and Pearce, B. M. (2003). "Team boundary issues across multiple global firms." *Journal of Management Information Systems*, 19(4): 157–190.

Galison, P. (1999). "Trading zone: Coordinating action and belief." In Biagioli, M., (Ed.), *The science studies reader* (pp. 137–160). London and New York: Routledge.

Fabbri, J. and Charue-Duboc, F. (2013). "Un modèle d'accompagnement entrepreneurial fondé sur des apprentissages au sein d'un collectif d'entrepreneurs: le cas de La Ruche." *Management International*, 17(3): 86–99.

Furnari, S. (2014). "Interstitial spaces: Micro-interaction settings and the genesis of new practices between institutional fields." *Academy of Management Review*, 39(4): 439–462.

Goglio-Primard, K. and Crepin-Mazet, F. (2011). "Organizing open innovation in networks- the role of boundary relations." *Management International*, 19: 135–147.

Hargadon, A. B. and Bechky, B. A. (2006). "When collections of creatives become creative collectives: A field study of problem solving at work." *Organization Science*, 17(4): 484–500.

Iaquinto, B., Ison, R. and Faggian, R. (2011). "Creating communities of practice: Scoping purposeful design." *Journal of Knowledge Management*, 15(1): 4–21.

Imbert, G. and Chauvet, V. (2013). "Faire coproduire le client en conception innovante. Les quatre processus mobilisés par les sociétés de conseil en innovation." *Revue française de gestion*, 39(234): 167–183.

Kritensen, T. (2004). "The physical context of creativity." *Creativity and Innovation Management*, 13(2): 89–96.

Levina, N. and Vaast, E. (2005). "The emergence of boundary spanning competence in practice: Implications for implementation and use of information systems." *MIS Quarterly*, 29(2): 335–363.

Lingo, E. L. and Mahony, S. O. (2010). "Nexus work: Brokerage on creative projects." *Administrative Science Quarterly*, 55(1): 47–81.

Menger, P. M. (2002). *Portrait de l'artiste en travailleur. Métamorphoses du capitalisme.* Paris, France: Seuil.

Mérindol, V., Bouquin, N., Versailles, D.W., Capdevila, I., Aubouin, N., Le Chaffotec, A., Chiovetta, A. and Voisin, T. (2016). *Le Livre Blanc des Open Labs. Quelles pratiques ? Quels changements en France ?*, Publication of the expert group facilitated by ANRT / FutuRIS PSB newPIC chair. Paris: ANRT et PSB (March).

Mustar, P. (1994). *Science et innovation 1995. Annuaire raisonné de la création d'entreprises par les chercheurs. Collection Innovation.* Paris: Economica.

Napier, N. K. and Nilsson, M. (2006). "The development of creative capabilities in and out of creative organizations: Three case studies." *Creativity and Innovation Management*, 15(3): 268–278.

Oakley, K. (2009). "The disappearing arts: Creativity and innovation after the creative industries." *International Journal of Cultural Policy*, 15(4): 403–413.

Penrose, E. (1959). *The theory of the growth of the firm*. Oxford: Oxford University Press.

Peredo, A. M. and Chrisman, J. J. (2006). "Toward a theory of community based entreprise." *Academy of Management Review*, 31(2): 309–328.

Pierre, X. and Burret, A. (2014). "Animateur d'espaces de coworking, un nouveau métier?" *Entreprendre and Innover*, 23(4): 20–30.

Schmidt, S. and Brinks, V. (2017). "Open creative labs: Spatial settings at the intersection of communities and organizations." *Creativity and Innovation Management*, 26(3): 291–299.

Sharma, P. and Chrisman, S. J. J. (2007). "Toward a reconciliation of the definitional issues in the field of corporate entrepreneurship." In: Cuervo, Á., Ribeiro, D., & Roig, S. (Eds.), *Entrepreneurship*. Berlin and Heidelberg: Springer: 83–103. The article has been reproduced from *Entrepreneurship Theory and Practice*, 1999, 23(3): 11–27.

Sternberg, R. J. (Ed.). (1999). *Handbook of creativity*. New York: Cambridge University Press.

Tushman, M. L. (1977). "Special boundary roles in the innovation process." *Administrative Science Quarterly*, 22(4): 587–605.

Verstraete, T. and Fayolle, A. (2005). "Paradigmes et entrepreneuriat." *Revue de l'Entrepreneuriat*, 4(1): 33–52.

Von Krogh, G. and Geilinger, N. (2014). "Knowledge creation in the eco-system: Research imperatives." *European Management Journal*, 32(1): 155–163.

Vyas, D., Heylen, D. and Nijholt, A. (2009). "Collaborative practices that support creativity in design." 11th European Conference on Computer Supported Cooperative Work, ECSW09: 151–170.

Wilson, G. and Herndl, C. G. (2007). "Boundary objects as rhetorical exigence: Knowledge mapping and interdisciplinary cooperation at the Los Alamos national laboratory." *Journal of Business and Technical Communication*, 21(2): 129–154.

Yin, R. K. (2009). *Case study research: Design and methods*. Los Angeles, CA and London: Sage Publications.

4 Living labs

New players in the dynamics of healthcare ecosystems of innovation

Alexandra Le Chaffotec and Valérie Mérindol

The goal of this chapter[1] is to appraise how living labs contribute to new dynamics of collaboration in healthcare ecosystems of innovation. Living labs represent original cases of open labs as defined by Mérindol and Versailles (2019). "A Living Lab is an open innovation research method aimed at the development of new products and services. The approach promotes a process of co-creation with end users in real conditions and is based on an ecosystem of public-private-citizen partnerships" (Living Lab White Paper, UMVELT, 2014). Their role relates to the development of the user-centric approach (Hakkarainen and Hyysalo, 2016; Schiavone, 2020) and complements the traditional approach based on a linear process of innovation. The experimentation process requires considering the spatial and organisational configuration in which a technological solution is tested. Many living labs offer a physical space and various tools for experimentation (Dube et al., 2014; Picard, 2017). Physical spaces play a central role in experimenting and enhancing collaboration in the user-centric approach (Oksanen and Ståhle, 2013; Said et al., 2017).

Nyström et al. (2014) show that the roles of living labs in ecosystems can vary a lot. Depending on the cases, they provide new services dedicated to the management of user-centric projects, orchestrate multi-partner collaboration, and/or design new solutions based on co-creation. In healthcare ecosystems, the variety of governance models used for living labs makes it complex to analyse the nature of their relations with the other actors in the ecosystem. Some living labs are not-for-profit organisations; others are private firms or units integrated in universities or hospitals. Many of them obtained the label issued by the European Network of Living Lab (ENoLL); others do not even seek to obtain this label.

Because of this heterogeneity, it is difficult to figure out how living labs contribute to the transformation of healthcare ecosystems of innovation. These ecosystems are characterised by a wide variety of interactions and collaborations among public and private actors that contribute to innovation processes (Pikkarainen et al., 2017; Dedehayir et al., 2018). Newcomers such as start-ups are also present in these ecosystems. Open innovation in healthcare implies the integration of a wide and diverse body of knowledge coming from heterogeneous actors such as entrepreneurs, students, designers, and users (patients and

DOI: 10.4324/9781003125587-7

healthcare professionals). To accelerate the adoption of innovation, multidisciplinary collaborations are required (Thakur et al., 2012; Kodoma, 2015). Collective ways of working in healthcare ecosystems of innovation must be transformed and adapted (Frow et al., 2016).

Various communities have emerged in healthcare ecosystems of innovation to contribute in overcoming organisational and cognitive boundaries (Pop et al., 2018). The development of these communities offers the opportunity to link a huge variety of competences. Communities are considered as the locus of the exploration process (Amin and Cohendet, 2004). They are characterised by a degree of self-organisation, shared values, and common passion. They are based on strong ties among individuals coming from various organisations and bound by the same practices. These communities contribute to tacit knowledge exchanges by developing informal relationships based on personal trust. They represent "shared spaces" (Creplet, 2000), where members share language, routines, and norms that allow deep learning and communication processes. They contribute to gather various profiles and to change the culture of innovation inside complex institutions. The open innovation literature underlines the variety of communities (West and Lakhani, 2008): communities of innovation, of experts, of practices, of users. Each type of community can be found within healthcare ecosystems. These communities allow dense exchanges of knowledge but their focus, goals, and compositions can differ a lot.

Even though living labs flourish in healthcare ecosystems of innovation, it remains difficult to understand how they contribute to their transformations and their complex dynamics of collaboration. To investigate how living labs contribute to ecosystems of innovation, this research is based on the analysis of 11 living labs located in France. Some 25 interviews were conducted between 2016 and 2021. Interviewees were mainly managers of the living labs and users involved in the living labs' activities (patients and professionals present in the ecosystems). We also interviewed entrepreneurs and students who participated in experiments. In line with Yin's recommendations (Yin, 2009), this research develops a multi-case investigation based on hypothesis-generating cases.

Our research shows that all living labs in the sample act as brokers of networks (Agogué et al., 2013). Brokers of networks contribute to the development of connections and enhance exchanges inside ecosystems. This chapter offers the opportunity to make this brokerage function more precise. In some cases, living labs reinforce knowledge exchanges in existing healthcare ecosystems; in other cases, they contribute in the generation of new dynamics in the ecosystems. The chapter identifies that living labs have three main contributions to the dynamics of collaboration inside healthcare ecosystems of innovation. Some living labs accumulate these roles. First, living labs contribute to the reinforcement of existing relationships among hospitals, private firms developing new technologies or new solutions, and academic institutions. In this first case, they intensify the interaction between public and private actors and facilitate the reference to the user-centric approach. Second, they contribute to developing new communities of practices with individuals who face the same complex problems. Third, they contribute to the development of communities of innovation. In this third modality, living labs directly contribute to connecting new actors together for co-creation activities.

This chapter also characterises how the living labs of the sample manage these three types of activities. The deployment of living labs implies dedicated competences to handle the experimentation of new solutions, their evaluation, and the management of collaborative projects. Living labs also devote lots of efforts to convince public and private actors to use their services. This research also shows that the management of the dynamics of ecosystems requires the management of the living lab physical spaces. It specifies the type of space required to encourage various forms of connections inside healthcare ecosystems of innovation.

The chapter is organised as follows: the first part in sections 4.1, 4.2 and 4.3 describe the functions of living labs that contribute to boosting ecosystems of innovation in the healthcare domain, first as a key contributor to fill the gaps in health ecosystems of innovation, then as an architect of communities of innovation and, finally, as a proponent of communities of practice. In each section, we then analyse the original modes of interaction implemented by living labs, managerial modalities applicable to these activities, and the design of the associated physical spaces. Section 4.4 discusses the results of this analysis and points out that most living labs cumulatively enact two out of these three roles. Section 4.5 concludes.

4.1 Living labs as key actors to bridge the gaps in existing ecosystems of innovation

4.1.1 New modes of interaction in established ecosystems

In the healthcare sector, one of the aims of living labs is to promote research orientations that are not addressed so far and to address gaps about everyday life issues that are important for users yet ignored by basic and clinical research. Living labs address new issues and promote new methodologies putting users at the centre of the innovation process. They make contributions to the development of new medical devices and have a specific focus on the adaptation of digital technologies. They also improve medical concepts by improving their adaptation to specific patients, family taking care of them, or healthcare practitioners. Living labs also appear as an additional building block active in the introduction of open innovation and user-centric innovation approaches into structured ecosystems of innovation (Nyström et al., 2014). All living labs investigated in this chapter match these observations.

Living labs are very often located within healthcare ecosystems, animated by hospitals or universities (see Table 4.1). These ecosystems are already organised but they require deep connections with technologists and new firms committed to the value chain of medical devices. Living labs support the intensification of these connections (Mérindol et al., in press): they organise events and seminars, support the experimentation phases, and/or provide tools for experimentation.

Table 4.1 Presentation of case studies

Living lab	*Date of creation and localisation*	*Governance of the living lab and label*	*Thematic orientation*
cLLAPS	2015 Located in Paris (installed in Pitié-Salpêtrière Hospital)	Foundation (not-for-profit organisation) owned by a medical and scientific institution (ICM) labelled by ENoLL	Imagines and proposes innovative solutions to meet the concrete needs of the care chain in neurology and psychiatry
Lab Santé Île-de-France	2016 Located In Paris (installed in Cochin Hospital)	Not-for-profit organisation. No label	Supports innovation in hospitals, nursing homes and clinics located in the Paris region with digital technologies
I-Care Cluster	2011 Located in Lyon. Installed in Tuba (a space that conducts collaborative innovation projects on the topic of smart cities)	Not-for-profit organisation labelled by ENoLL	Cluster in Rhone-Alpes Auvergne region. Specialised in digital technologies for healthcare and med tech
HealthCare Factory – Hacking Heath Strasbourg	2012 Located in Strasbourg	Private firm. No Label	Organisation of experimentation of new technological solutions for hospitals, nursing homes and clinics
CEREMH	2008 Located within Paris-Saclay University and Moveo (mobility valley ecosystem)	Not-for-profit organisation labelled by ENoLL	Design and deployment of innovative solutions promoting mobility for people with disabilities

ActiveAgeing	2014 Located in Troyes, in Troyes Technology University	University department Labelled by ENoLL	Develop and test support solutions for elderly people
Allegro	2018 Angers, in the city hospital	Hospital department and expertise centre labelled by ENoLL	Co-design and evaluation of services and new technologies for gerontology
La Fabrique de L'Hospitalité	2017 Located in Strasbourg, in Strasbourg University Hospital	Hospital department labelled by ENoLL	Improve working conditions of hospital practitioners, and the care of patients and relatives, through design methodologies
Lusage	2010 Located Paris, in Broca Hospital specialised in gerontology	Public research laboratory labelled by ENoLL	Evaluation, co-design, and development of technological solutions for the quality of life of elderly people (with or without mental disabilities) and their supporting staff or families
Streetlab	2011 Located in Paris, in Institut de la Vision, inside Hôpital des Quinze-Vingts (specialised in ophthalmology)	Private firm linked to the hospital and to medical institutes. Labelled by ENoLL	Co-design and evaluation of solutions to improve the autonomy, mobility, and quality of life of visually impaired people
CIC-IT Lille	2008 Located in Lille, in Lille University Hospital	Public research laboratory labelled by ENoLL	Support for innovation, co-design, evaluation of e-health, and biosensor solutions; support to the of innovative technologies in healthcare in general

During experimentation, living labs gather different stakeholders who already know each other but need new exploration methods. They contribute, therefore, to the dynamism of local ecosystems in healthcare with new services such as the evaluation of medical devices, methods for the experimentation of new technological solutions, and support during the operation of experiments. They operate mainly at the request of actors of the ecosystem who need support to evaluate a medical device or to experiment.

Lab Santé Île-de-France is located in the Paris region. It is active in innovation projects during the definition of research topics and for the design of experimentation in public hospitals and private clinics; it also supports the identification of partners appropriate for each experimentation project. To developers of medical devices, it proposes the right people with the adequate level of seniority and influence in order to make the experimentation of new solutions easier. Lab Santé Île-de-France acts as a guide in the experimentation process. It convinces professionals in healthcare systems to test and use new solutions. It also helps stakeholders to find public subsidies for the experimentation.

> We have a fairly exhaustive vision of the needs of healthcare actors and understand the impact of potential innovations. We know how to put these people in relation with each other for the benefit of the general public healthcare system.
>
> (Manager of Lab Santé Île-de-France)

Streetlab is a living lab located in the Vision Institute (Institut de la Vision) in Paris. It is governed by a public-private partnership where Quinze-Vingts Hospital plays a major role. Streetlab combines basic, clinical, and industrial research on the same site. Both institutions are leaders in France in the field of ophthalmology. Streetlab's mission is to improve the autonomy, mobility, and quality of life of visually impaired people. The portfolio of activities includes co-design, evaluation, rehabilitation, and awareness raising. The living lab conducts projects that are not traditional clinical or basic research projects (doctors would not have time to work about subjects such as lighting and marking, even though these issues make the daily life of visually impaired people much easier) but which are no less useful and complement them. It offers public and private actors (municipalities, airports, etc.) new solutions to facilitate the mobility of visually impaired people. It co-designs with digitalisation actors daily life apps adapted for visually impaired people. It evaluates industrial projects (for example augmented-reality glasses). For all these activities, Streetlab works with multidisciplinary teams, bringing together physicians with industrial companies or start-ups. Most of the time, Streetlab follows the methodology of clinical trials and benefits from the database of users it shares with Quinze-Vingts Hospital. To carry out these services, Streetlab uses technological platforms that reproduce elements of daily life (a flat, an artificial street, a car simulator).

Centre de Ressources et d'innovation Mobilité Handicap (CEREMH) has its headquarters in the south of the Paris region and operates as a not-for-profit organisation. It supports mobility stakeholders to promote solutions for people with reduced mobility or disabilities. The living lab works closely with the "Pôle de compétitivité" Moveo, an industrial and R&D cluster specialising in urban mobility, and with Paris-Saclay University. It has, for instance, developed various tools (simulator, training programme) for driving schools, health professionals (ergo-therapists), and end users. These activities of the living lab, in turn, allow professionals to provide services dedicated to disabled people (issuing driving licences for disabled people, training sessions, etc.). CEREMH's approach is to offer new and innovative services that complement technological solutions provided by scientists and engineers for people with reduced mobility.

> While there are many things available in the field of disability in France, there is often a disconnection, between what can be offered and the real needs of people with reduced mobility. This is particularly true in the world of research and innovation.
>
> (Manager of CEREMH)

La Fabrique de l'Hospitalité is a unit of Strasbourg University Hospital. This case perfectly illustrates how a living lab provides a new component of services in an existing ecosystem. La Fabrique de l'Hospitalité proposes new categories of services around the reorganisation of medical services with tools such as design, signage, etc. This service aims at improving the reception of patients and families, and the working environment of the hospital staff. Users are involved in the projects from the writing of the specifications to the delivery of the output.

Similarly, CIC-IT Lille is a research unit of Lille University Hospital. It contributes to developing new approaches of innovation in existing ecosystems. It specialises in e-healthcare, in biosensors, and in the use of innovative technologies for healthcare. This lab supports clinicians, academics, manufacturers, and users in the design, evaluation, and valorisation activities of innovation processes. The main mission of this living lab is to facilitate the access to the hospital technical platform to evaluate medical devices on behalf of the industry or of other research laboratories. Through its evaluation, CIC-IT ensures that the medical devices under development meet specific clinical safety requirements. To carry out tests and evaluate the quality of medical solutions, the living lab makes it possible to reproduce the conditions of operating rooms and collect data about body temperature, blood pressure, etc. or operate electrocardiograms. This information is then digitalised, allowing for data analysis and rapid prototyping. In this context, a start-up based in Lille has, for instance, developed a device measuring in real time the pain of sleeping patients during surgery, based on the analysis of heart beats. This instrument allows the surgeon, who visually controls the screen, to adapt his or her action or the dose of analgesic or anaesthetic products according to the degree of pain felt by the sleeping patient. This technology is now used in different European hospitals.

4.1.2 Managerial modes to offer new services in existing ecosystems with living labs

To strengthen the ties between members of existing healthcare ecosystems of innovation, field research shows that living labs require two important properties: individual competencies to align ways of working between contributing innovation stakeholders, and a space where they operate together and therefore contribute to strengthen these ties.

To spur innovation based on the user-centric approach in the (local) healthcare ecosystem, living labs teams need to understand the key challenges, interests, and constraints of the ecosystem. They also need to host the appropriate skills and competencies to support experimentation and evaluate innovative solutions. The network brokerage requires the development of complementary skills for the team operating the living lab. These people must understand the various interests and constraints representing each actor in the ecosystem. The facet of the network brokerage function mobilised here does not so much relate to connecting people together and to creating new links; it focuses on enabling people from different backgrounds to collaborate and combine their respective approaches to facilitate the emergence of an innovative solution. Living lab teams must translate the different languages inherent in each field of expertise to other contributors. They improve, for instance, the potential for communication between companies in the digital world and hospitals. Because living labs understand the dynamics of existing ecosystems, they can communicate with all the stakeholders, to understand each category of stakeholder, and to connect public and private actors around experimentation.

The multidisciplinary nature of the open lab team is also essential to carry out a constant monitoring and analysis of the users' needs, which makes it possible to design projects adapted to these multifaceted needs. Depending on the skills required by the living lab's activities, its team can be made up of people with medical or engineering backgrounds, or people with experience in the administration of healthcare institutions, or with ergo-therapists, psychologists, designers, etc. The understanding of needs also stems from a good knowledge of all difficulties encountered by users targeted by living labs, and from a good analysis of their environment(s).

All living labs studied in this research operate with an emblematic space in a specific location: most of them have hosted their facilities within hospitals (cLLAPS, Allegro, StreetLab, Lab Santé Île-de-France, CIC-IT Lille, La Fabrique de l'Hospitalité, Lusage). Some of them are hosted by a university (CEREMH, ActiveAgeing) or have selected a location in a place that already carries similar values and facilitates similar projects: I-Care Cluster is based at Tuba, an iconic space in the French city of Lyon, which runs collaborative innovation projects on the theme of smart cities.

To develop this new category of services inside existing ecosystems, living labs implement a functional space adhering to descriptors introduced by Oskanen and Ståhle (2013): related spaces exhibit properties that are keys to managing projects based on the user-centric approach. In many cases, these spaces are equipped with technology and tools for experimentation. They are therefore used for the materialisation of the original services provided by living labs: the space becomes a platform for experimentation. The space itself can be defined as a service, since it can be used on behalf of a third party which conducts experimentation on the premises of the living lab: the living lab makes the space itself and its technical features available as a service, and eventually complements it with supporting competencies. Spaces are usually flexible and adaptable, and hence tailored to each innovation project. If the living lab's space has several rooms, the team will select the one best suited to the proposed activity (creativity workshop, experimentation, seminar, etc.). Spaces also contribute to the development of new modes of interaction with the end users (healthcare practitioners, patients, and their relatives, all users of innovative solutions). This represents one of the challenges in the user-centric approach. The space is proactively designed and managed to host various activities. Its design looks for reconfigurability. In several cases investigated in this research, the physical space is also a temporary area suited for user-centric experimentation, configured *ad hoc* for specific projects. Lab Santé Île-de-France, for instance, sets up its activities directly inside the hospital standard premises to meet healthcare practitioners where they work. The I-Care Cluster also installs experiments in specific micro-spaces in partner hospitals.

4.2 Living labs as architects of new communities of innovation

In healthcare ecosystems, the emergence of new communities of innovation represents another challenge when they are made up of various profiles such as executives from large companies, entrepreneurs, technological geeks, designers, patients, healthcare practitioners, or scientists. The notion of a community of innovation was defined in the introduction. These communities gather people who have not interacted together previously but who share the same motivation to contribute to the transformation of the healthcare innovation system by actively participating in the development of new solutions.

4.2.1 New modes of interaction to enhance communities of innovation

This research shows two main issues in relation with the development of communities of innovation in healthcare systems. First, it is necessary to create a climate of trust to enforce knowledge exchanges and support the development of creative ideas among people who do not know each other. Second, it is necessary to help people develop new collective practices based on the adoption of a user-centric perspective to solve problems and elaborate on fast prototyping.

Four living labs in the sample contribute to the emergence of communities. Their main activity is to host events to encourage people who have never previously worked together to connect and identify collaboration opportunities. They also organise hackathons, facilitate the use of innovative user-centric protocols such as design thinking, ethnographic approaches, and collectively contribute to problem resolution.

cLLAPS is the living lab of the "Brain and Spinal Cord Research Institute" (Institut du Cerveau et de la Moelle Épinière) (ICM). ICM wants to create a new ecosystem to develop solutions for patients suffering from diseases in this broad research domain. These solutions can be based on new treatments, e-health projects, or med techs. In parallel to its cLLAPS living lab, ICM has set up an incubator, and launched technological platforms with adapted services to accelerate solutions provided by the private sector. All activities contribute to the emergence of new dynamics for this community of innovation. As per its main mission, cLLAPS co-conceives the experimentation of new technological solutions in the hospital with new actors such as start-ups and students. It develops initiatives to encourage the adoption of user-centric innovation approaches. Various profiles are present: in 2021, hackathons have gathered students coming from schools of engineers and schools of design, start-ups hosted in the ICM incubator, healthcare professionals, and representatives from patients' associations. These people had never met before the hackathon. Multiple consequences of this event ensued. The hackathon created numerous temporary connections and helped people to work differently. "The bet was that students will be more empathetic with future users after being in contact with patients" (Manager of cLLAPS). Some of these connections outlasted the hackathon. Small teams (many of them the winners of awards received during the hackathon) decided to continue their project and develop new innovative services to people with Parkinsonism. They were supported by cLLAPS to go deeper into the exploration, for instance with coaching to explore technological feasibility and then experiment with the solution in a real environment.

Activities are quite similar with the I-Care Cluster (located in the Rhone-Alpes-Auvergne area), and with the Health Care Factory located in Strasbourg. These two organisations host Hacking Health regional chapters and aim at introducing exploratory processes based on the living lab approach. Hacking Health is an international network of people passionate about the transformation of healthcare systems (see Chapter 5). They offer a methodology to encourage the emergence of new communities after events (bootcamps and hackathons), and rapid experimentation of new solutions. I-Care Cluster and Health Care Factory follow Hacking Health's protocols. The challenge is to gather various competences to "*hack the healthcare systems*". Events offer opportunities to create new connections and help people develop new collective creative practices. Hackathons gather more than 500 people on a yearly basis. They trigger positive experiences for the healthcare professionals who contribute and realise that they can create solutions and improve their daily routines.

> The impact of the event is mainly due to the acculturation of people. People need to understand the impact of digital solutions on healthcare. We are waiting for individuals to gain the hacker mindset and to understand the relevance of hijacking a technology to solve a concrete problem. We want them to understand that they have now a methodology to invent something new.
>
> (Manager of Health Care Factory, in charge of the Hacking Health Strasbourg chapter)

4.2.2 Managerial modes to develop communities of innovation with living labs

Contributing to the development of communities of innovation implies specific modes of management based on hosting events and managing physical spaces to encourage new connections. These modes of management are quite different from the modes of management adopted by the living labs to contribute to exchanges within existing ecosystems.

Lab Santé Île-de-France reveals specific combinations of events and meetings to create communities of innovation. This living lab has launched two main initiatives. First, it identified the most interesting digital entrepreneurs for healthcare systems located in the Paris region. With them, it animates networking activities to share key issues about healthcare. The living lab connects and aligns visions and interests between start-ups and institutions such as nursing homes (EHPAD in French), or public and private hospitals located in the Paris region. It organises meetings and supports start-ups to translate healthcare professionals' expectations into e-services. The second initiative focuses on the management of new explorative projects developed with multiple partners such as local policymakers, bio-scientists, sociologists, designers, hospitals, entrepreneurs, insurance companies, healthcare professionals, or urban service mobility to explore the future of healthcare services. This initiative is built in collaboration with Liberté Living Lab, an open lab dedicated to new technologies, civic techs, and public utilities (see Chapter 1). The Liberté Living Lab team has developed specific competencies for the animation of creative seminars in such multi-partner contexts.

These living labs have put the animation of events at the centre of their initiatives to encourage the emergence of communities of innovation.

In the four cases promoting the emergence of communities of innovation, managers support the adoption of new creative methodologies in their respective networks. They show empathy and curiosity for testing new ideas and gather various creative people. In these cases, interviewed managers explain again that physical spaces play a major role in supporting this emergence. However, originalities prevail. Conversely to living labs focusing their intervention on the rejuvenation of existing innovation ecosystems with functional spaces, living labs fostering the emergence of communities usually install temporary spaces. All living labs described in Section 4.2 use temporary spaces: partners such as universities, hospitals, or other open labs offer, for instance, access to spaces to organise hackathons or various seminars. These spaces are adapted and reconfigured to install the conditions for successful creativity sessions. They also carry a symbolic content (Oskanen and Ståhle, 2013) and embody values and symbols in relation to the visibility that living labs want to diffuse and promote in their respective ecosystems. In the list of living labs supporting the development of communities of innovation, only I-Care Cluster works in parallel with a permanent space and with temporary installations. I-Care Cluster uses its permanent space as a sort of totem place embodying the values of openness and collaboration. Thanks to the design of this space, I-Care Cluster manages focus groups and seminars and helps people adopt a creative posture. Larger events are then organised in dedicated temporary spaces.

The four living labs analysed in this section exhibit properties identified for interstitial spaces, as described by Furnari (2014) and Vilani and Phillips (2020). They contribute "to gather people coming from different institutional fields to discuss and experiment with matters of common interest" (Vilani and Phillips, 2020: 2). They provide a favourable physical context to develop creative and informal exchanges among people who do not know each other. Interstitial spaces play a role as drivers of new interactions and support the transformation of collective practices into innovation. They are managed adequately to produce new ideas and practices: spaces are organised to make exchanges easier, to facilitate a climate of trust, and generate serendipitous encounters. Interstitial spaces must be considered as a neutral arena for exchanges that are materially adequate to contribute to complex problem resolution. The property of neutrality for these arenas means that these spaces do not support a single vision of innovation and that living labs are not pushing forward any particular interest (either related to public or private stakeholders present in healthcare ecosystems). Physical spaces help people adapt their respective posture. "One of the advantages of our partner [*Liberté Living Lab*] is that they have a welcoming physical space that is completely outside the rivalries and institutional constraints existing in healthcare" (Manager of Lab Santé Île-de-France). The analysis of these four cases shows how their managers pay careful attention to the design and the selection of these spaces, both when they work with permanent and temporary spaces selected *ad hoc* for a particular event (such as a hackathon).

4.3 Living labs as proponents of communities of practice

4.3.1 New modes of interaction to encourage the development of communities of practice

Living labs nurture the development of communities of practice (Wenger, 2000). These communities are formed by individuals who share the same practices and/or have the same problems to solve. They communicate with each other about their respective knowledge areas. A strong pattern always prevails among members of these communities: they want to invent local solutions to problems encountered in their professional or personal practices. When these people interact, share, and disseminate their knowledge and expertise, they learn together and create new knowledge. Their relationships are based on mutual commitments, and they share common resources such as language, tools, routines, methods, etc. In this context, they also discuss best practices, and benefit mutually from each other's expertise, for instance on how to solve a problem that one of them has already encountered. In healthcare ecosystems, individuals who are part of these communities may be healthcare practitioners or end users (patients and relatives). Other professionals such as entrepreneurs can share knowledge and experience on a variety of topics (who to contact, how to finance a project, what to know about business creation, etc.). Patients are more likely to exchange tips and tricks about diseases, treatments, practitioners to consult, or possible approaches to manage a given disorder. Mutual assistance is at the heart of the community's operating mode. Its goal is to improve and share the best methods or practices.

Relationships remain informal, horizontal (no hierarchy: all individuals are considered as peers), and most often temporary. Interactions take place inside the living lab during seminars, coffee breaks, etc. However, it remains complex to maintain relationships outside the living lab and over long-term timeframes. The formation of communities of practice emerges step by step, with meetings and repeated interactions, within the framework of activities proposed by a living lab. Events, seminars, and informal breaks during experimentation hosted by the living labs all represent the triggers for the dynamics of communities of practices.

4.3.2 Managerial modes to develop communities of practice with living labs

To encourage the implementation of communities of practice, living labs organise several types of activities or events in their facilities. The idea is to set up meetings with a certain regularity, installing the dynamics of interaction between participants thanks to their recurrence. Living labs organise sessions that give time for exchange between peers, foster learning dynamics, and allow feedback from individual experience. The facilitation of workshops requires specific profiles within the team managing the living labs. Facilitators must be recognised actors in the ecosystems. Their legitimacy is based on experience in the field where knowledge exchanges take place. They must also be empathetic with end users, which is especially important in the healthcare domain for patients and elderly people: they need to be reassured, convinced about the relevance of contributing to such activities (debates, experiments), but also made aware that they own experiential knowledge accurate for new innovative practices. Four living labs in our sample enact this role inside healthcare ecosystems of innovation.

ActiveAgeing organises "thematic breakfasts" with end users to work with co-design methods. These workshops are organised around a theme that will be discussed by the different participants. The intervention of the facilitation team is important to start the debate and to raise different questions around the theme under discussion. The team facilitates interactions and is directly at the origin of the density of exchanges. Over the years, ActiveAgeing has progressively developed a community of more than 300 people, mostly patients who are more than 60 years old, interested in working on autonomy issues and now empowered to share with providers of new solutions. These users have now become the experts of their own needs in the co-creation process and the living lab can regularly reach out to them to test new medical devices.

Lusage organises thematic workshops in its "café multimedia" in similar ways. This living lab is located in Broca geriatric hospital in Paris, of which it is a unit. It focuses on geriatric problems and cognitive impairments, such as Alzheimer's disease. It develops and evaluates digital solutions such as software, applications, robots, etc. for elderly people. Workshops organised by Lusage represent an opportunity to discuss modern techniques and digital solutions addressing dependence issues for seniors and people with mental disabilities. They are also an opportunity to collectively put into practice and use these digital tools, test them, discuss their usefulness, their usability and adaptation to seniors, and introduce corrections. Users find it very interesting to see other people using these tools rather than considering them alone. They learn more collectively. Moreover, debates are also enriching the innovation process because they share opinions and ideas that might not have emerged if they were alone. "I love hearing Mr [*other user of the living lab*] develop his ideas. You can hear that he went to university while I didn't" (user from Lusage).

CEREMH encourages communities to develop thanks to a digital platform that it has installed. Users with disability and mobility issues are encouraged to discuss their views on discussion forums. They exchange advice, discuss opinions, and, also, formulate needs that the living lab can identify, thus influencing its research and development projects. This community is original because it only operates with digital interaction tools.

Allegro regularly organises idea generation workshops with practitioners from the geriatric department of Angers hospital. These workshops are based on face-to-face interactions and therefore bring together caregivers, physicians, or nurses. Events aim to exchange views on medical practices dedicated to patients hosted in the geriatric department and to generate ideas that patient follow-up.

There are some limitations to building these communities of practice, especially when it comes to end users. The first limitation relates to reasons given by users who contribute to such workshops. Their motivations are usually rather self-oriented. They want to understand technologies or inventions that might ease their own situations and address their own problems in the future, and make sure that these solutions are consistent with their own perspective of ethics, or privacy (which represent a sensitive issue when dealing with digital solutions). "I am from the generation that began to experience technical progress and will need technology in my third and fourth ages" (a patient active in Lusage). A second limitation relates to the durability of these communities of practice. Users interviewed in our sample indicated that they did not meet or exchange outside the activities proposed by the living lab. These communities are therefore not self-organising as it is usually the case with communities (Wenger, 2000). "The people I meet at the multimedia café, I don't see them again afterwards" (user from Lusage). Ties between people do not seem to be as strong as advocated by the literature (Amin and Cohendet, 2004).

The last comments about the management of living labs in relation to communities of practices develop again about the issue of the physical space, which plays an important role in their emergence. The reference to Oskanen and Ståhle (2013) was already introduced in previous sections about managerial actions in relation to the physical space. Again, their research is relevant here to describe how living labs impact, generate, and improve the dynamics of communities of practice: spaces must be "attracting" and "value reflecting". However, once again, originalities prevail because of the specificities of the orientation towards the promotion of communities of practices. Communities of innovation usually elaborate on professionals from different sectors (not only healthcare), with significant expertise in their respective fields, but communities of practices are (also) populated with users who locate their expertise in their actual experience, including those of patients or of patient's families. The management of living labs aiming at communities of practices has therefore to attract people who do not own pre-specified knowledge. A major condition to enhance interactions remains the easy accessibility of their spaces, to make sure that all people who feel that they have something to say are easily welcomed. It is therefore important to work without ID or badge screening at the door, and to select locations that are easily accessible by public transportation or to ensure the availability of parking places nearby the living lab, etc. In a nutshell, the management of the physical spaces and of the related events is not targeting the same objectives. When linked with communities of innovation, the living lab managers design the space to make sure that community members overcome their own individual limitations and contribute together to innovation projects. When linked with communities of practice, the living lab managers work to attract "users" and new profiles into the living lab; they ensure that the design of the space is suited to motivate contributions by community members who do not have a prior understanding of innovation processes.

In this context, as with interstitial spaces described in Section 4.2, physical spaces aiming at promoting communities of practice must be user-friendly and designed to make sure that all categories of "users" feel comfortable, want to come back, and become loyal to the community of practice. As already described in the previous sections, physical spaces are often designed to offer a cosy design and welcoming atmospheres. However, in contrast to the previous category, these spaces are not primarily designed to foster creativity, they are intended to stimulate communication and exchange. They are designed to promote easy access and easy interactions. Lusage has, for instance, installed a round room with a round table to facilitate communication and exchange between users participating in workshops and experiments. By seeing each other, users are more likely to talk to each other, to observe each other's experience, which in turn enriches the collective experience in the community of practice. "Where we meet is round. [...] When it is round, everyone is at the same distance" (user from Lusage). This circular disposition of the physical space remains a specificity of the cases contributing to the elaboration of communities of practice, because users who accept to play a recurring role in these communities have an even more significant need than participants in communities of innovation to feel comforted in a reassuring context. It is important for living lab managers to offer a fertile space disposition for exchanges, with special attention paid to end users. From the sample of living labs investigated here, it seems that "user-friendliness" is more important for physical spaces supporting communities of practice than for the ones supporting communities of innovation. In both cases, living lab managers manage the attractiveness of their physical spaces and organise regular events, and contribute to strengthen links between participants. To generate or facilitate the dynamics of their respective communities (communities of innovation versus communities of practice), they foster different drivers of conviviality and contribute to develop a climate of trust in pursuing different goals. In communities of practices, they work for participants to feel comfortable and thus fertilise interactions, provide recurrent contributions, and most notably overcome their heterogeneity (in particular, the heterogeneity between their different statuses, or between the discrepant origins of their respective "expertise"). The effectiveness of a community of practice in healthcare will increase when all categories of members, including healthcare practitioners, patients, and (eventually) families will not be shy to contribute to the different activities. The design of the space plays both a functional and symbolic role in this respect.

4.4 Discussion

This research identifies three strategic options for living labs in healthcare ecosystems. Option 1: they energise healthcare ecosystems by creating new bricks of innovation and promoting new methods based on the user-centric approach of innovation. Option 2: they allow the creation of communities of innovation. Option 3: they encourage the creation of communities of practice. From the sample of living labs investigated in this chapter, we identify that each option requires a specific mode of management. Different skills and competencies are required for each option. This research also shows that the living lab's physical space always represents a key managerial tool. However, the characteristics of the space are specific to each modality, to match the main drivers and activities prevailing for each strategic option. It is also interesting to analyse that living labs have sometimes chosen to combine some of these options, and therefore also mix the subsequent managerial modes, the associated modalities, and the properties of the physical space.

4.4.1 A combined contribution of living labs to the dynamics of ecosystems

This chapter identifies three common contributions by living labs. First, they always combine several functions. Second, they contribute to the adoption of a new vision of innovation based on the user-centric approach in healthcare ecosystems that were initially framed by a predominant culture of top-down and science-push innovation processes. Third, they contribute to the development of connections and exchanges in evolving ecosystems.

In our sample, living labs always bridge the gaps in existing healthcare innovation ecosystems by promoting new methods based on the user-centric approach. It is possible to characterise this contribution further in pointing out that they have a focus on the development of different communities, therefore the three options identified at the beginning of this section. In option 1, they only fill innovation gaps in ecosystems. In option 2, they fill these gaps while being architects of new communities of innovation. In option 3, they fill these gaps while promoting communities of practice. Table 4.2 summarises this result. Only three cases in our sample solely focus on strengthening the dynamics of

Table 4.2 Synthesis of the contributions by living labs to healthcare ecosystems of innovation

Contributions	*Competencies and skills*	*Design of the space*	*Cases in the sample*
Option 1. Reinforcing ecosystems	People who master experimentation and evaluation methods. Capacity to enable people with different backgrounds to collaborate. Multidisciplinary monitoring and analysis of the users' needs	Functional space with technical devices to carry out experiments	• La Fabrique de L'hospitalité • Streetlab • CIC-IT Lille • CEREMH • ActiveAgeing • Allegro • Lusage • cLLAPS • Lab Santé Île-de-France • I-Care Cluster • Health Care Factory – Hacking Heath Strasbourg
Option 2. Creating communities of innovation	Capacity to create new links. Competencies for the facilitation of seminars with new creative methods. Empathy and curiosity	Interstitial space to allow for creativity, informal exchanges, and serendipity. Neutral and temporary space. Space is a totem place embodying the values of openness and collaboration	• cLLAPS • Lab Santé Île-de-France • I-Care Cluster • Health Care Factory – Hacking Heath Strasbourg
Option 3. Boosting communities of practices	Legitimacy, empathy. Ability to convince people to contribute to workshops. Capacity to improve the density of interactions during workshops	User-friendly space with easy access, offering a welcoming environment and a cosy design to encourage participants' loyalty	• CEREMH • ActiveAgeing • Allegro • Lusage

their ecosystem (option 1). All other living labs combine option 1 with one of the other options. However, we did not find any situation where a living lab drives both types of communities at the same time (and therefore mixes options 1, 2, and 3 together). We have a potential explanation for this dichotomy: the modes of management of these communities and the required skills are not, together, consistent for options 2 and 3. The cases show, for instance, that the management of events to stimulate the creation of communities of innovation, such as hackathons, is so specific that it is often necessary for the living lab team to be supported by specialists of the facilitation of these events, and that all living labs do not engage in the organisation of such activities. In contrast, skills required to bring new elements to innovation ecosystems, and the subsequent management of the living lab physical space, are easy complements either to those of an architect for communities of innovation or those of the promoter of a community of practice.

To strengthen links between partners in the ecosystem, living labs promote experimentation methods. They can understand the various challenges of their ecosystem and help people to work together; they energise relationships inside their ecosystem. They also contribute to translate meanings and to share a common perspective. These properties characterise living labs as "brokers of networks" (Agogue et al., 2013). As architects of communities of innovation, living labs go further into this brokerage function when they manage different types of events to encourage co-creation activities among people with different backgrounds. When living labs promote or support the development of communities of practice, the brokerage function reaches out to people who often also participate in the experiments (patients in particular).

All living labs act as brokers of networks whatever their purpose or the "option" listed in Table 4.2. To fulfil this brokerage function, their teams act as boundary spanners (Levina and Vaast, 2005). In the three options described in the table, it is necessary to understand how to interact with the different categories of stakeholders involved (or targeted by) the living lab activities, and to enable all their representatives to communicate and to work together. As boundary spanners, living lab teams contribute in translating meanings and create a common ground (i.e., a common understanding of the problems to be solved) among heterogeneous actors (for instance, technology geeks, physicians, technologists, end users, designers, sociologists, etc.). Boundary spanners translate language and build bridges to enable all these worlds to meet. It contributes to the alignment of interests and visions. Hence, living lab teams often have a multidisciplinary composition. The brokerage and boundary-spanning functions are not ensured by a single person in these teams but operated instead by all complementary skills present in the living team. Lab Santé Île-de-France is a perfect example when they describe themselves this multidisciplinary effort jointly delivered by their team of six people.

> We all come from different backgrounds. [...] The director of Lab Santé Île-de-France is a professor specialised in cardiology [...]. He brings

> medical credibility. [*As deputy director of this living lab,*] I have accumulated experience in social insurance administrations and regional healthcare agencies. I bring knowledge about political, administrative and financial issues. In the team, we have also [...] dual-education profiles: engineers [*from different specialisations*] and an MBA graduate specialised in healthcare management [*who*] holds the position of referent for start-ups.
>
> (Deputy Director of the Lab Santé Île-de-France)

The organisation and animation of events represents a core aspect of the dynamics impulse by living labs, contributing to the emergence of both types of communities. This may help to explain why communities have short lives, even sometimes an ephemeral existence, because they depend on the moments shared with events, seminars, hackathons, etc. In this perspective, living labs are the trigger for new networks of practice: they contribute to the emergence of temporary informal relationships and to the emergence of common interests based on one-off interactions. By introducing regular and multiple events and seminars, most living labs expect to transform these networks of practice into actual communities of practice where strong ties and self-organisation prevail.

For living labs, the challenge is therefore to build these communities and encourage the sustainability of connections, or the transformation of weak and temporary ties (network of practice) into strong ties (communities of practice). In this perspective, the regularity of events and the setting-up of a symbolic space for interactions and collaboration are key instruments. The living lab works as a catalyst for the emergence of communities. Through the repetition of events, it expects to create changes in individual and collective practices, and a more sustainable dynamic of interaction.

4.4.2 The central role played by the living labs space

This research shows that physical spaces play a critical role in living labs. All previous sections of this chapter identified links with the roles and properties mentioned by Oskanen and Ståhle (2013), as it is consistent with the user-centric approach to innovation. Physical spaces support collaboration and promote interactions. The cases offer opportunities to clarify two points. First, that the main functions and properties of the space vary according to the contribution of the living lab to the dynamics of their ecosystem. Second, that a living lab can manage several (types of) spaces to fulfil various functions inside their ecosystem.

We have identified that the attributes of the physical space differ with the mission, the orientation being either functional, or interstitial, or user-friendly (see Figure 4.1). The physical space represents a managerial tool for the living labs, a means to an end. Depending on the mission of the living lab, spaces will play a different role: a service, a communication tool, a convivial space, etc. Living labs adapt their space to the mission they fulfil and to the public they

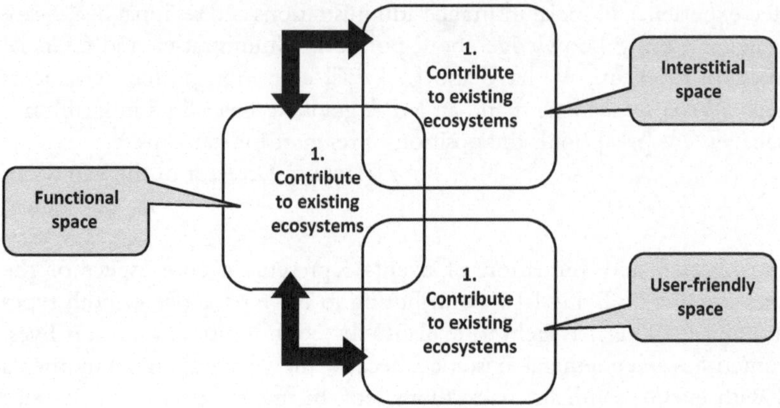

Figure 4.1 Key roles for physical spaces in the strategies of living labs.

address. Most living labs have installed multipurpose physical spaces, reconfigurable and flexible. Living labs pursuing several missions at the same time (rejuvenation of innovation ecosystems, promotion of communities of innovation or of communities of practice) have therefore the option to best adapt to the nature of activities, or events, and take advantage of the different technological or conviviality properties. In specific cases, living labs have targeted dedicated fractions of their facilities to technical projects and experimentation, while other areas support the lives and activities of their communities. Adaptability and reconfigurability always prevail. All cases show a direct practical link between the multipurpose nature of the space and the ability to perform several types of functions in the innovation process (support, co-design, evaluation, community enhancing, etc.). Preserving these aspects obviously generates new adaptation challenges when aiming at installing or accelerating innovation in the healthcare sector with digital technologies.

Figure 4.1 shows that living labs that perform the mission of energising existing ecosystems of innovation (option 1) mobilise a functional (i.e., technical) space. This space can be modified to carry out the mission of the architect of communities of innovation (option 2), and it then becomes an interstitial space with a temporary configuration conducive to creativity. The living lab can also develop the characteristics of the space adapted to promote communities of practice (option 3) with a more convivial and welcoming space (therefore less technical, for example).

The cases show the existence of the combination of options 1 and 2, or of options 1 and 3, therefore the arrows on the figure. The combination of options 2 and 3 was not present in the sample. This means that living labs may elaborate specific strategies based on physical spaces that can be considered as functional and interstitial at the same time, or that can be jointly characterised as functional and "user friendly".

No specific reason may explain why the combination of a functional, interstitial, and user-friendly space makes sense to elaborate a living lab strategy, but we suggest that the reason for this incompatibility may be located at several levels: either in the conditions appropriate to generate communities of innovation versus communities of practice, or in the timeline of activities developed by the community working in the living lab, or in the conditions for the emergence and operation of the physical space as a boundary space (Champenois and Etzkowits, 2018). This issue represents perspectives for further research.

4.5 Conclusion

Living labs support the development of a user-centric approach in healthcare ecosystems of innovation. This process involves different actions such as the creation, test and evaluation of innovative solutions, and the originality of living labs elaborates on the strong commitment of end users in some (or all) related activities. Living labs have been traditionally considered as new actors that bring tools and methods for experimentation. This chapter shows that their contribution goes beyond the mere technical aspects of experimentation. In this context, new collective contributions are required for the dynamics of innovation inside ecosystems. This chapter goes deeper into this analysis with a focus on healthcare ecosystems. In the sample investigated in this chapter, living labs contribute to the dynamics of innovation by supporting the creation of new communities (of innovation, or of practice) and by enhancing their operations.

The development of communities (of innovation and of practice) represents a huge challenge. Living labs use different tools and perform different roles to stimulate exchanges and connections in healthcare ecosystems. The space plays a key role in this process that will foster future ideas and innovation. The living labs do not only animate events and other activities when encouraging the emergence of communities of innovation and of practice. They also consolidate and sustain their existence and activities in a long-term perspective.

Note

1 Many thanks to David W. Versailles for helpful comments and insightful suggestions during the development of this contribution.

Bibliography

Agogué, M., Yström, A. and Le Masson, P. (2013). Rethinking the role of intermediaries as an architect of collective exploration and creation of knowledge in open innovation. *International Journal of Innovation Management*, 17(2): 1–24.

Amin, A. and Cohendet, P. (2004). *Architectures of knowledge: Firms, capabilities, and communities*. Oxford: Oxford University Press.

Champenois, C. and Etzkowitz, H. (2018). From boundary line to boundary space: The creation of hybrid organizations as a triple helix micro-foundation. *Technovation*, 76–77: 28–39.

Creplet, F. (2000). The concept of "ba": A new path in the study of knowledge in firms. *European Journal of Economic and Social Systems*, 14(4): 365–379.

Dedehayir, O., Mäkinen, S. J. and Roland Ortt, J. (2018). Roles during innovation ecosystem genesis: A literature review. *Technological Forecasting and Social Change*, 136: 18–29.

Dube, P., Sarraih, J., Grillet, C., Zingraff, V., & Kosteck, I. (2014). *Le livre blanc des livings labs*. Montréal: EMVELT.

Frow, P., McColl-Kennedy, J. R. and Payne, A. (2016). Co-creation practices: Their role in shaping a health care ecosystem. *Industrial Marketing Management*, 56: 24–39.

Furnari, S. (2014). Interstitial spaces: Microinteraction settings and the genesis of new practices Between institutional fields. *Academic Management Review*, 39(4): 439–462.

Hakkarainen, L., & Hyysalo, S. (2016). The evolution of intermediary activities: Broadening the concept of facilitation in living labs. *Technology Innovation Management Review*, 6(1), 45–58.

Kodoma, M. (2015). *Collaborative innovation: Developing health support ecosystems*. New York: Routledge.

Levina, N. and Vaast, E. (2005). The emergence of boundary spanning competence in practice: Implications for implementation and use of information systems. *MIS Quarterly*, 29(2): 335–363.

Merindol, V. and Versailles, D. W. (2019). *« Créer et Innover aujourd'hui en France et en Asie : le rôle des plateformes d'innovation et des open labs d'entreprises »*. Research co-funded by Innovation Factory and Bpifrance le Lab. http://innovasia.newpic.fr.

Merindol, V., Versailles, D. W. and Le Chaffotec, A. (2019). *Répondre aux défis du management de l'innovation en santé. Le rôle des dispositifs d'intermédiation en France*. Research funded by Genopole, Evry: France. http://innov-sante.newpic.fr.

Merindol, V., Versailles, D. W. and Le Chaffotec, A. (2021). Les organisations intermédiaires et l'innovation : les multiples facettes de l'intermédiation de réseau. *Innovations*, 2(65): 49–80.

Nyström, A. G., Leminen, S., Westerlund, M. and Kortelainen, M. (2014). Actor roles and role patterns influencing innovation in living labs. *Industrial Marketing Management*, 43(3): 483–495.

Oksanen, K. and Ståhle, P. (2013). Physical environment as a source for innovation: Investigating the attributes of innovative space. *Journal of Knowledge Management*, 17(6): 815–827.

Picard, R. (2017). *Co-design in living labs for healthcare and independent living*. London: Wiley.

Pikkarainen, M., Ervasti, M., Hurmelinna-Laukkanen, P. and Nätti, S. (2017). Orchestration roles to facilitate networked innovation in a healthcare ecosystem. *Technology and Innovation Management Review*, 7(8): 30–43.

Pop, O. M., Leroi-Wereds, S., Roijakkers, N. and Andreassen, T. W. (2018). Institutional types and institutional change in healthcare ecosystems. *Journal of Service Management*, 29(4): 593–614.

Saidi, T., De Villiers, K. and Douglas, T. S. (2017). The sociology of space as a catalyst for innovation in the health sector. *Social Science and Medicine*, 180: 6–44.

Schiavone, F. (2020). *User innovation in healthcare how patients and caregivers react creatively to illness*. Springer Nature. https://doi.org/10.1007/978-3-030-44256-9.

Thakur, R., Hsu, S. H. Y. and Fontenot, G. (2012). Innovation in healthcare: Issues and future trends. *Journal of Business Research*, 65(4): 562–569.

UMVELT. (2014). *Le Livre Blanc des Living Labs*. Montreal. https://www.montreal-invivo.com.

Villani, E. and Phillips, N. (2020). Formal organizations and interstitial spaces: Catalysts, complexity, and the initiation of cross-field collaboration. *Strategic Organization*, 19(1): 5–36.

Wenger, E. (2000). Communities of practice and social learning systems. *Organization*, 7(2): 225–246.

West, J. and Lakhani, K. R. (2008). Getting clear about communities in open innovation. *Industry and Innovation*, 15(2): 223–231.

Yin, R. K. (2009). *Case study research: Design and methods* (4th ed.). Thousand Oaks, CA: Sage.

5 From spatiality to temporality

Turbocharging innovation ecosystems with events: the case of Hacking Health

Luc Sirois and Karl-Emanuel Dionne

The orchestration of ecosystems to drive innovation is increasingly important for business and society. Addressing contemporary business and societal challenges now relies on a variety of stakeholders coming from different domains, organisations, and sectors to drive the development of ground-breaking innovations built on complex value propositions (Adner, 2006; 2017; Walrave et al., 2018). It is not surprising then that theories on innovation ecosystems have increasingly attracted researchers' and managers' attention (Phillips and Ritala, 2019).

An innovation ecosystem is the evolving constellation of heterogeneous actors, activities, and artefacts, and the institutions and relations that are important for the creation of value (Adner and Kapoor, 2010; Autio and Thomas, 2014; Granstran and Holgersson, 2020). The innovation ecosystem framework helps focus on the co-evolutionary processes taking place as various actors, from different domains and organisations, interact to create and deliver value (Dedehayir et al., 2018). These different actors can include producers, suppliers, distributors, financial and research institutions, makers of complementary technologies, and regulatory bodies. This definition helps accentuate innovation ecosystems as (temporally) evolving phenomena constituted by the connective ties between actors and activities from different institutions. Actors in innovation ecosystems look to collectively create, deliver, and appropriate value (Nambisan and Sawhney, 2011), which opens considerable issues related to how and when they come together to create and deliver such value. Moreover, within ecosystems, the creation and development of innovation communities is an important issue, as they represent collectives within ecosystems, the smaller cells of the overall organisms. However, while innovation ecosystems are difficult to orchestrate given their broader and constant evolution, innovation communities can be used as mechanisms to assemble and funnel large groups of people and organisations around a shared value proposition and shared means of producing and delivering such value.

One of the main challenges that have been explored by researchers interested in innovation ecosystems and communities relates to their spatial dimension, and how innovation can be orchestrated when potential contributors do not meet in their natural work environments. One way to solve this puzzle has

DOI: 10.4324/9781003125587-8

been the development of creative spaces such as open labs – which are spaces specialised in finding and developing creative solutions from the combination of actors from different domains and organisations (Mérindol and Versailles, 2017). While early scholars of ecosystems have identified temporal dynamics as central elements to be considered in understanding the functioning of ecosystems (Aarikka-Stenroos and Ritala, 2017; Moore, 1993), existing answers to ecosystem orchestration have led to an overemphasis on strategies and research on these "lieux" where people from different organisations can coalesce to develop ties, share knowledge, and participate in innovation activities. However, this focus on the spatial dynamics involved in driving collaboration has somewhat ignored the temporal dynamics at play in innovation ecosystems. Consequently, we know little about the temporal processes involved in innovation ecosystems: what sparks the emergence of collaborations and fosters their growth, what makes innovation communities form and help them grow in size and productivity within ecosystems, what triggers and sustains their innovation activities, what creates their stability or dynamic nature. We know little about the sequencing of such processes, how and when they occur, and how to improve their impact. This represents major barriers to advancing our collective understanding of how innovation ecosystems form and function, and how they make innovation communities emerge and grow over time, how to accelerate their progress. In this chapter, we argue that temporal dimensions are as important as their spatial counterpart in such complex and continuously evolving systems. Temporality is essential to fully comprehend, and eventually influence, the development and orchestration of ecosystems.

In this chapter, we build and expand on a literature in organisational theory that has a focus on temporal dynamics at play in collaboration and innovation activities within organisations. We use this literature to highlight temporal differences across stakeholders' domains, organisations, and institutions. This literature has depicted how organisational actors craft temporal niches where change can occur outside existing temporal constraints (Fine, 1990), negotiate their different temporal regimes (Kaplan and Orlikowski, 2013) and create synchronicity across their differences (Granqvist and Gustafsson, 2016). One could also consider the temporal dimensions that arise from "interaction" across these interdependent actors and the inherently dynamic nature of innovation ecosystems, which evolve over time. We also aim to help open lab managers and community leaders integrate them as powerful tools in their innovation leader arsenal.

In this chapter, we expose how events play an essential role in managing the temporal dynamics of innovation ecosystems and their subcomponents, such as innovation communities. Events can bolster the development and efficiency of innovation processes and communities (Lampel and Meyer, 2008; Cohendet et al., 2014). Events are essential tools orchestrators can use in connection with collaborative spaces to temporally orchestrate ecosystem dynamics. Without events, they risk only creating, at best, physical places where people and ideas meet, without consideration for all the other dynamics at play. Without events, their open labs and creative spaces can be like hardware without software, bodies without souls. Without events, they will fail to create and nurture the necessary communities and underlying dynamics for their open labs to deliver results, for innovation ecosystems to exist and thrive.

5.1 Methodological issues and field research: the Hacking Health case study

In this research project, we used a revelatory case study (Corley and Gioia, 2004; Eisenhardt, 1989; Yin, 2009) to document how Hacking Health, a non-profit organisation focused on growing innovation, used events to shape the temporal dynamics of unstructured digital health ecosystems. Authors have either been part of or studied the organisation from its outset and have been involved in organising events to orchestrate these ecosystems. We use this insider view to describe Hacking Health's strategic and intentional use of events in driving ecosystem development.

Different aspects of the Hacking Health approach will serve to illustrate the concepts highlighted in this chapter. Hacking Health is a "global movement to improve healthcare", with active chapters in 41 cities in 16 countries. It is a highly decentralised non-profit organisation, a kind of "league" of community leaders sharing the goal to spark, accelerate, and sustain innovation in the healthcare sector, all in their own regions scattered around the world. They organise in "Hacking Health Chapters", which are groups based in cities or small regions around the world. Their leaders are very often head of open labs of all sorts – innovation hubs, incubators, accelerators, innovation centres – in their hospitals and cities. They deploy inspired efforts to facilitate the creation of concrete solutions to real-world health problems. Ultimately, they aim to make local healthcare organisations more innovative and agile. In their quest, they deploy an arsenal of approaches from the open lab toolbox. Not only do they create a global movement and give more visibility to their initiatives worldwide, but also rapidly develop the art and science of events as a powerful tool to accelerate innovation and community development. Using Hacking Health as a case study, we expose how events help engage key stakeholders across domains, organisations and sectors, bolster creativity, generate new projects, and create momentum. We explore why events play a central role in the open lab framework, and how they can be powerful tools for innovation ecosystems to generate lasting impact.

5.2 The five essential functions of events for ecosystems

Our research and years of orchestrating innovation ecosystems with events with Hacking Health and other organisations, shows that events play five essential functions for ecosystems: (1) magnets, (2) mines and mixers, (3) momentum makers, (4) metronomes, and (5) markers. As this section explains these five functions and highlights various temporal dynamics at play in innovation ecosystems, it aims to expose the different roles events can play in temporally orchestrating innovation ecosystems. We show that despite their seemingly temporary impacts, events can drive lasting developments by impacting the temporal structures that are essential for our understanding of innovation ecosystems given the way they drive momentum, connect people from different domains, organisations, and sectors and spur collaborations among them. By doing so, we expand on a mostly spatialised understanding of the dynamics of innovation ecosystems.

Firstly, events are *magnets* as they attract key players, help connect them together, and can hold communities together. Secondly, events are *mines and mixers*. They are fertile grounds for mining people's brains for new insights. They can boost creativity by mixing ideas to generate new ones, by establishing new contacts, and finding original solutions. Thirdly, events are *momentum makers*. They fuel human energy. They can help people engage more actively and persevere in the innovation process, instil discipline and a sense of urgency, increase their pace. Fourthly, events are *metronomes*. They create a shared tempo, a common rhythm across different domains, organisations, and sectors that would otherwise evolve at their own individual pace. Finally, events are *markers*. They crystalise, or "mark", how far communities of innovation have gone, who they are, what they know, and what they could achieve at a point in time. For open labs, they are milestones along their journey, the stepping-stone by which to measure progress and to leap on the road ahead.

5.3 Events

But first, what type of events are we talking about?

5.3.1 "Traditional events"

What comes to mind when mentioning events are often traditional formats such as conferences, symposiums, seminars, conventions, corporate offsites, and board meetings. A literature on field-configuring events (FCE) – defined as temporary social organisations that assemble diverse members of an organisational field in a bounded time and space to exchange information or coordinate activities (Lampel and Meyer, 2008) – has emerged to better understand the impact of temporary events on fields and ecosystems (Jolly and Raven, 2016). Most empirical settings upon which the FCE literature was developed involve conferences, trade fairs, award shows and other highly institutionalised "megaevent" formats, such as the Grammy Awards and trade fairs. The images of speeches or presentations in front of an audience easily come to mind. Less formal alternatives can also come to mind and still be considered part of the more traditional mix: team meetings, project reviews, town hall meetings, corporate offsites, or even social or networking events. Traditional events are moments, places, and structure that bring actors together to share knowledge, results, progress, status, and news (Lampel and Meyer, 2008). While these "traditional events" are less interactive and action-oriented than the "creative events" described below (Citroni, 2015), they are equally important in the temporal dynamics of innovation ecosystems and communities.

5.3.2 "Creative events"

Creative events are interesting contexts to look at attempts to gather a variety of actors and resources to generate new technologies and spark ecosystem change. Events such as innovation contests (Boudreau et al., 2011), design competitions (Lampel et al., 2012), and hackathons (Briscoe, 2014) are designed to attract a greater variety of actors to foster the development of communities and innovation (e.g., Boudreau et al., 2011; Bullinger et al., 2010; Almirall et al., 2014; Briscoe, 2014). Creative events can be contest-based and are designed to foster contributions from a variety of actors despite the actors' commitments to other paths of activities. These events can funnel actors' interests towards the focal area's specific needs and challenges (Bullinger et al., 2010; Murray et al., 2012; Johnson et Robinson, 2014). Such creative events provide times and places where actors come together to reflect, interact, generate new ideas, construct new concepts, generate new designs, work on projects, all in a very hands-on and interactive manner. Such "creative events" are a place for more hands on deck, a moment for very hands-on experiences that deliver results. They are events where participants actually create something together, actively and interactively building on each other's knowledge, insights, and contribution. Creative events could range from simple brainstorms to more structured workshops, from design jams to hackathons, to more elaborate design sprints such as the ones made popular by Google in the start-up community, and even the large-scale, month-long, to "cooperathons" invented by Hacking Health and Desjardins in Quebec. Creative events bring together actors of a particular pursuit. During creative events people go beyond sharing knowledge. They interact to generate new knowledge. Creative events aim to achieve results, to help participants build something together. The time and place are set, but special attention is dedicated to putting participants in a particular state of mind, to fuel their creative juices, to create a time-space outside. They do so through a variety of means and approaches from format, facilitation, and setting but also with artefacts or artistic performances targeted at stimulating the right side of the brain.

> Hacking Health uses events "to connect, engage, educate, collaborate, and inspire. Our chapters create moments where they bring together individuals with diverse expertise and a collective will to have an impact on healthcare. Hacking Health Cafés [*traditional events*] provide a friendly environment where members of the community can share their experience, have open discussions with mentors and coaches, and network with other members. Creativity is fed by knowledge and insights. Hacking Health cafés are gatherings for the community to connect and learn. Technology experts learn about healthcare's challenges and opportunities.

Clinicians learn about the realm of possibilities brought to life by technology. Startups showcase bold ventures and visions of the future".

(HH website, 2020)

Hacking Health uses various creative event formats to drive creativity and promote the healthcare innovation agenda. Spreading over different periods of time, they vary in scale and in the diversity of participants they attract. "Hackathons are Hacking Health's most famous [creative] events. For 48 hours, individuals with diverse expertise are brought together to build meaningful solutions to realworld health problems. Teams design and prototype solutions to pitch a jury of experts" (HH website, 2020). Beyond hackathons, other creative events are deployed depending on the amount of time and resources available, as well as the desired impact (Figure 5.1).

For Hacking Health, hackathons and all their variations – from design jams to a hackathon blitz, to innovation challenges and the more elaborate cooperathons – are some of the most effective ways to bring all stakeholders together and begin a quest towards innovation. Their events are designed to free up participants' creativity, spark new ideas, amplify their individual capabilities, and develop their cross-domain collaborative abilities. Hacking Health aims to raise the bar and catapult events in general, and creative events in particular, to a new level of impact on innovation ecosystems and communities.

5.4 Events as magnets

Just like magnets, events have the power of attraction. They act as magnets of attention, magnets of people, magnets of communities. Many mechanisms are at play behind this phenomenon.

Events are magnets of attention. Events have, by definition, been moments in time, a fixed duration with a beginning, and most importantly an end.

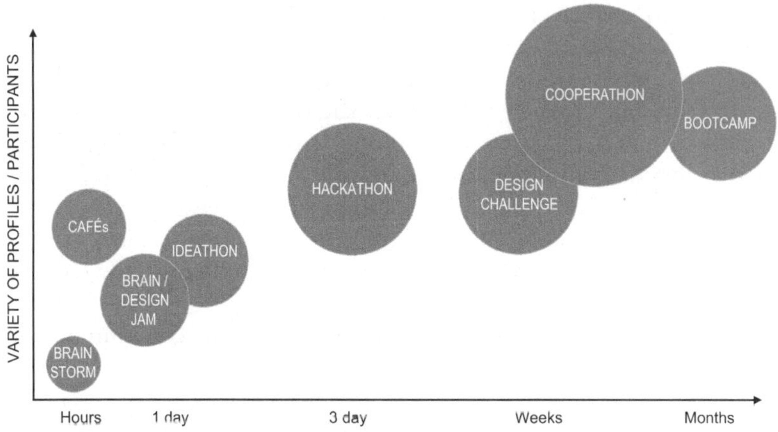

Figure 5.1 Types of creative events used by Hacking Health.

They represent a "now or never" situation. Their temporary nature and infrequency make them stand out as "newsworthy", or worthy to be talked about, worthy of attention. And indeed, events command attention. Because of this, and because they are outside of regular routines, they "hit the radar". They become more visible "blips" moving closer on the screen. They become "worthy" of being communicated about, and that, in turn, makes them stand out even more. Even while they are happening or after they concluded, events are sources of newsworthy reports. Open labs can use them to draw attention to an agenda, a mission, an organisation.

Events are magnets of people. People are drawn towards events for a variety of reasons, from simple curiosity to fear of missing out, from a desire to learn, to live an experience, or to connect, events draw the attention of many potential participants. Some will see in events an opportunity to share their learnings, others to advance their personal or corporate pursuit. In all cases, events tap on the fundamental gregarious nature of human beings who intrinsically look for opportunities to come together. Wise open lab leaders will use them to attract and engage key stakeholders.

Events are magnets for communities. Magnets to form them, magnets to keep them together. Events attract and bring together actors of a community, but also, as the Hacking Health history exposes, they can help *create* communities. Past the initial events, a string of future events will help community members stick together and attract others, helping the community grow or span out in a greater variety of community interests. If properly managed, open lab managers and innovation community leaders have at hand a powerful tool to create a virtuous circle. Events, particularly if they are repeated on a regular basis, hold people together, make their community more attractive, and help grow it over time. Sequences and strings of events will be addressed below.

Hacking Health organisers quickly realised how events were powerful ways to trigger attention, attract people, and create innovation communities in healthcare. For them, building communities first is essential for successful high-profile creative events like hackathons but at the same time, high-profile creative events like hackathons provide a powerful driving force to build communities. In other words, in the Hacking Health approach, building communities is not only seen as a necessary *first step* towards using events as drivers of innovation, but also a *consequence* of using events to drive innovation. Their focus on communities creates a virtuous circle of mutually reinforcing mechanisms, with high-profile creative events acting as magnets in both directions.

The Hacking Health approach prescribes building communities by reaching out and connecting with appropriate stakeholders, getting them intrigued by the challenge at stake, and getting them acquainted and interested in each other. It achieves this purpose by organising "Hacking Health Cafés", traditional events combining formal presentations with ample interactions and an important networking component. These "cafés" bring together the individuals that will eventually be called upon to take action. In all cities and regions, Hacking Health systematically organises such events hand in hand with established players and community leaders. "Together, as complementary actors, we achieve stronger results. We're here to energize partners with new tools, resulting in modern approaches to each partner's own missions". In all cases, Hacking Health use events as magnets to bring together actors from a variety of domains.

- **Healthcare.** By design, Hacking Health events are intended to attract patients, healthcare professionals, physicians, and decisionmakers from hospitals, academic medical centres, faculties of medicine, clinics, pharmacies, as well as various associations and foundations. They attract anyone from the curious to the driven changemakers who aspire to improve things around them, frontline actors who see first-hand the real problems to solve and hope something could be done about them. Hacking Health events are magnets to them as they offer unique opportunities for them to expose their perspective on the issues to solve, as well as their ideas on how to solve them. As good innovations often stem from solving real problems, this is invaluable in the Hacking Health innovation process.
- **Technology.** While patients and healthcare professionals have strong ideas on the problems they want to solve and the processes and barriers they face that need to be changed to do so, they typically lack the required technology-related knowledge on how to build the solutions. This is where technologists and designers kick in. Hackathons are natural magnets to software designers and engineers who are attracted by technical challenges and the opportunity to show off their technical skills. The difference with Hacking Health events is that it provides them with an opportunity to do so for a higher purpose, for a noble cause. Hacking Health hackathons cater to the generations of professionals in search of purpose and meaning in their lives and are powerful magnets to those who want to deploy their skills and talents to make a difference in the world.
- **Start-ups.** Entrepreneurs and staff from start-ups, accelerators, coworking spaces – typically from the digital health sector – naturally gravitate towards Hacking Health events for many reasons. It could be to connect with their end users, to accelerate the development of components of their solutions, find new ideas, recruit contributors, or find new sources of help. Such events are also positive ways to gain exposure to decisionmakers from the corporate, government, investment, and healthcare domains.
- **Government.** Representatives from local, municipal, state, provincial, and federal bodies, economic development agencies, as well as grant-making agencies are attracted to the Hacking Health events to connect with their constituencies, learn about new ways to achieve

their mission and to grow their impact. Over time, top government officials often use their presence at the high-profile Hacking Health events to demonstrate or exercise public leadership. They in turn provide legitimacy to the events and act as magnets for other participants.

- **Academia and research.** Undergraduate and graduate students have always been strong adepts and participants of hackathons. They fuel project teams with their skills, energy, and motivation to show what they are capable of. More and more, researchers and professors from universities, research centres, and consortia are attracted to hackathons and other creative events as new ways to accelerate and increase the relevance of their research. Creative events also provide academic actors new ways to explore industry collaboration, tech transfer, and translation opportunities.
- **Corporations.** Representatives of the corporate world are important ingredients in the mix. Typically coming from the digital health, medical technology, pharmaceutical, or insurance sectors, enterprises can facilitate the implementation and scaling-up of creative ideas developed in the hackathons. Their staff can provide valuable advice and insight to hackathon participants. But they also come to learn how to accelerate the pace of innovation, the level of energy and motivation within their own organisation, and to connect with the creative "underground" communities of innovators (Cohendet et al., 2014).
- **Investors.** Some of the corporate players are attracted to Hacking Health events from an investor's perspective, but the "investors" category stands on its own as it is mainly composed of venture capital firms. Hackathons and other entrepreneurial community events are natural magnets for venture capitalists (who sometimes participate in the organisation of such events themselves) who often contribute to their organisation. What makes the Hacking Health events particularly attractive to them is the opportunity to mingle with a broad variety of actors interested in their domain, to see the early seeds of what could become real start-ups, and to discover eventual investment opportunities in new areas of potential growth.

As can be seen from the motivations of many of the participants described above, creative events such as the Hacking Health hackathons go beyond being magnets. They are "mines and mixers".

5.5 Events as mines and mixers

Places – innovation hubs, creative office spaces, incubators, etc. – promote the importance of coming together. They are sometimes perceived as the only natural canvas to serendipity, lucky encounters, tacit knowledge. These phenomena also happen, if not more so, at events. Within a short and fixed period of time, events can actually amplify these phenomena with more people, more intersections, more opportunities for these phenomena to materialise.

Events create ephemeral spaces. They also promote the importance of coming together, beyond existing sectorial, organisational, and disciplinary boundaries. Events help make knowledge come alive and ideas emerge.

Events are mines and mixers of ideas, people, problems, and solutions. Events truly are fertile ground for mining people's brains, intuitions, tap on their knowledge and skills, mine for new ideas, insights, and connections. Creative events go further and purposely boost the creative process. Not only do they make the best ideas emerge, but they also make them collide, mix, and intertwine to generate new ones. They catalyse collisions of ideas and people, connect problems and solutions, fuse competencies across disciplines.

5.5.1 Hacking Health events as mines

Hacking Health events illustrate particularly well how events can be mines of solutions and ideas. Firstly, Hacking Health hackathons, as opposed to what typical hackathons do, are always held with and around end users: healthcare professionals, patients, caregivers, and maybe system administrators. They always start with real-world challenges of people on the frontline. Technologies in search of problems to solve are well-known failures. Technology push must be avoided at all costs, particularly in the very hard and demanding healthcare "markets". On the opposite side, necessity is the mother of invention. Great innovations start with great problems, which is exactly the way Hacking Health events are structured.

Moreover, Hacking Health events provide a sandbox where participants can experiment as much as they want, with no real consequences whatsoever should they make mistakes, without anyone's boss watching or judging them. A liberating space of freedom to mine intuitions, hunches, aspirations, and perspectives from a variety of contributors. And this is important in the mining process as, like Jeff Bezos once said, "failure and innovation are inseparable twins". Yet, the fear of failing still prevails among innovation ecosystem stakeholders, despite all the advice to the contrary and the motto repeating "every failure is a step to success". Only one new product idea in thousands will become commercial successes, only one step in hundreds will actually be conclusive. Knowing this truism, Hacking Health organisers prepare participants to try a lot, to take many shots like in any sport. Whether by planting many seeds, by making many attempts, running various experiments in parallel, or by creating ways to fail fast, great innovators must play the odds. They must find ways to have the resources and resilience to try lots of different things, to mine lots of different sites if they wish to ever strike gold.

Drawing from the Hacking Health case study, we can find for instance that Hacking Health Cafés are specifically tailored to travel and be hosted in different areas of an ecosystem to mine for new potential people, their problems, and ideas. They travel from place to place, from month to month, specifically to mine in a variety of places and organisations and attract new people in their innovation community. Hacking Health workshops, on the other hand, mine for ideas. They help participants dig deeper in their experience, analyse the problems they face, identify the barriers to their progress or opportunities to cease, in a way to generate better-defined ideas. If shared in Hacking Health hackathons, these ideas can be further explored, refined, and turned into prototypes. Hackathons themselves are powerful opportunities to mine for people, technical skills, knowledge, and resources.

5.5.2 Hacking Health events as mixers

Hacking Health events also illustrate particularly well how events are mixers. First because they use the "wisdom of the crowd". Yet, acting as magnets, they are designed to assemble actors from different fields, backgrounds, skills, and competencies, as creativity lies at the intersection of them all. But this common wisdom is further developed by Hacking Health. Crowds, mentors, experts, and participants are called upon to provide perspectives and insights on a broad number of questions. The mass of participants brings about a deep set of experiences and backgrounds. They could, and are, used to test hypotheses, validate approaches, and prototype, *during* the event, and at any time.

Hacking Health is often seen as pairing healthcare professionals with technologists – software developers, engineers, etc. But Hacking Health organisers deploy real efforts so that designers are always included. As much as possible, elements of design are injected at all steps of the creative events. Beyond this trio of healthcare professionals, technologists, and designers, if the project was not carried by patients at the onset, then Hacking Health organisers look for ways to make sure patients are in the mix. They always aim to include patients in any project team. "Nothing about me without me", patient advocacy groups claim (Puckrein, 2016), and from a healthcare creative process point of view, Hacking Health organisers believe they could never be more right. As challenging as it may be, including the ultimate end user in project teams is a source of constant validation and insight.

Finally, beyond mixing participants, talents, and ideas, Hacking Health most often also tries to mix event organisers. They join forces with other organisations to broaden the impact and appeal of their mutual events. They look for synergies and avoid duplicating efforts. They do so by either adding the "innovation" and "creative" dimensions to popular pre-existing "healthcare" related events, or by adding the "healthcare" dimension to popular pre-existing "tech" events. As a result, together they act as magnets and mixers of greater crowds and move each other's agendas further by having a stronger effect on the temporal dynamics of their ecosystems by tying multiple innovation communities together.

In all cases, a temporality factor comes into play. Events can act as mixers at the right moment: at the beginning of the idea generation process for example, users, patients, or designers are more important. They help foster a common understanding between all participants. Events must bring together the right mix of people at the right time. Moreover, mixing must happen for the right amount of time. Not all participants are always required in a project. The involvement of some participants might be required for certain project phases but not for others, or only for a certain period. Using events to time the moment and duration of key stakeholders' involvement is an approach well understood by Hacking Health.

5.6 Events as momentum makers

Events have the particular power to fuel human energy, helping people engage and persevere in the innovation process. Most importantly, they instil discipline and a sense of urgency that makes ideas emerge. As a result, they increase the rate of development of innovation processes and ecosystem development. In other words, events can make people and communities move, and fast.

Actors of innovation, particularly in decentralised or "loose" structures but even in tighter, more traditional organisations, can benefit from setting a point in time for their deliverables. Imagine a special task force on a mission, with all members aiming to achieve a particular result at a particular point in time. On such a time-bound assignment, everyone aims to achieve their part of the project with a common deadline in mind. The work of each team member might evolve in a different direction and at a different pace, but with the shared understanding that all deliverables will have to be delivered at a particular point in time. Events can naturally act as such moments in time. The artful manager understands the power of events as "bookshelves" of innovation processes: their start and finish lines. The beginnings and end of the journey. Astute leaders purposely create events to provide a target to their team, something to aim for, something to work towards. Events, therefore, provide discipline to move forward. It creates a context to measure progress according to set objectives, enabling us to break down the different deliverables and expectations in smaller chunks of time. Even more importantly, the "in-person" nature of events, even on a virtual basis, bolsters all these attributes. Events require team members to show up in front of their peers and colleagues, in front of an interested audience, to present the fruit of their efforts. This creates even more discipline-inducing pressure.

5.6.1 "No battle plan survives contact with the enemy"

In businesses and organisations, managers often get busy defining well-structured action plans, with clockwork precision and clear marching orders to achieve predictable progress over time and to walk towards the ultimate deadline with a predictable linear pace.

And then reality hits.

As the above saying goes, no strategy can fully anticipate all aspects of on-the-ground execution. Linear project plans rarely deliver on their promises. Progress often starts slow, teams encounter difficulties. People get organised, ideas emerge, and unforeseen problems get in the way. Contexts evolve, lessons are learned. Plans were designed on misled hypotheses. The thing is, discipline does not always come naturally, or come at all. Team dynamics might get in the way. Personal skills might not be up to par. As a result, progress is slower than expected, at best. Projects can get stuck altogether, or worse, actually regress. Or the project teams might wander and explore – either by choice or inadvertently – and take paths totally different than initially planned, getting to totally different outcomes than anticipated. In all cases, linear project planning has many shortcomings in creative endeavours. So, if linear progress is rarely achieved in creative processes, do events as deadlines really matter? They do (see Figure 5.2).

Events increase the rate of development of innovation processes and ecosystem development. They can instil discipline and focus for project teams and communities of practitioners alike. In the face of difficulties and delays, having set a target in time forces teams to catch up as the deadline gets closer. It is a well-known phenomenon: progress accelerates as we get closer to the deadline. The approaching event will make the team members rally, prioritise, focus, look for solutions, and even increase their capacity to solve problems. Together they increase their pace of work. They work harder. They accelerate the pace. All for the sake of achieving the target. And if worse comes to worst, even if the target is not met, significant progress will be achieved under the pressure of a looming event, clearly more than would have been achieved otherwise.

Hacking Health events themselves impose incredible time limits. This pressure fuels the creative juices of participants. It forces them to always focus on priorities, achieve everything using the 80/20 rule. This is particularly true in hackathons which last only 48 hours. The same principle applies in longer marathon types of creative events, like the design challenges, cooperathons, or bootcamps, during which teams will work on their progress for days, weeks, or months, but always with a set deadline, and always with the same discipline

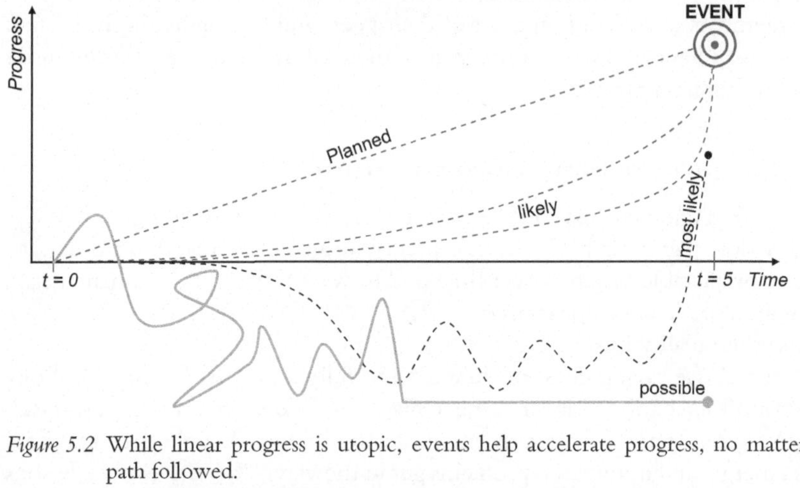

Figure 5.2 While linear progress is utopic, events help accelerate progress, no matter the path followed.

imposing time constraints. Events also "de-risk" this process by providing opportunities for teams to test new ideas without the pressure of having to succeed. Events force a deadline without necessarily being a deadline of having to produce the intended outcomes by that time. Events, therefore, are a fertile ground for trial-and-error processes, which in the end can accelerate the process of mining for the right way to solve a problem after a continuous process of failing fast.

The drastic timeboxing imposed by Hacking Health has another benefit: whereas in real-life people throw good money after bad at mediocre projects, at Hacking Health events, bad projects will face a hard stop without unduly dragging efforts and resources for more than two days.

5.6.2 More action, less talk

Hacking Health events are momentum makers also because they focus on taking concrete steps and actions towards particular goals. By definition, all Hacking Health events are action-oriented by design. But even the more subdued ones, like cafés and information sessions incorporate at least some activities enabling participants to get engaged, to actually do something, from as simple as expressing opinions, sending messages on social media, making new connections, to drafting complete manifestos, white papers. Even in "traditional events", Hacking Health organisers always scratch their heads to include elements that will induce change, to engage participants. They make time in the programme for speakers and the audience to make plans together, and even take the first steps towards achieving it, right there *during* the event. This gets the participant to activate and creates momentum. Hacking Health organisers often include a competition and award prizes in their events. For some reason, even for the most restrained participants, calling on humans' competitive nature, in a fun and benevolent way, always seems to improve their pace and quality of work. In Hacking Health events and others, people always "up their game" knowing they will have their work presented, evaluated, and "compete" for a prize. This energises participants and creates momentum.

5.7 Events as metronomes

Events create a shared tempo, a common pace across different domains, organisations, and sectors. Each of these collectives operates at its own pace, and without events, they do not have the necessary conditions to converge around shared temporal perspectives. Understanding the above dynamics, leaders of all sorts started to use events as "metronomes" by creating a series of events. With a series of targets and missions to accomplish at regular points in time, one can harness the power of events to fuel the momentum of innovation teams and whole ecosystems alike. Metronomes are devices that create a drumbeat, keep musicians playing alone on a regular cadence, and help all musicians of a group stay on the same rhythm. The concept of metronomes in an open lab and innovation ecosystems setting introduces notions of regularity, coordination, and synchronisation. In the open lab context, events become devices that produce regularly repeated spaces. They push practitioners and innovators to work at a particular speed. They give them opportunities to synchronise efforts across different organisational and epistemic rhythms.

5.7.1 Metronomes to build momentum and keep going

Events create momentum. But as the early Hacking Health leaders rapidly learned, one event alone will not *maintain* the momentum. In the Montréal chapter first, and in all cities and chapters after, the Hacking Health experience demonstrated how events held on a regular basis help build, grow, and maintain momentum and communities (see Figure 5.3).

In Montréal, a group of visionaries decided to organise a first hackathon. These co-founders of Hacking Health, later joined by other co-founders from many cities, jumped right into organising what became a surprisingly successful creative event. Physicians and people from all over the country flew in to participate. As did politicians, investors, and corporations. The co-founders saw they had touched a nerve – a deep desire among healthcare professionals and technology specialists to create solutions to improve healthcare. They went on and organised similar events in other cities across the country, far from Montréal.

During this time, enthusiasm rapidly dwindled at home. The success of the first event had been misleading. While it was more than just a flash in the pan as it became the spark of what is now a global movement and attracted the attention of critical stakeholders, it failed to create long-lasting engagement among participants. It was a memorable moment that acted as a magnet, a mine, and a mixer to temporarily spur creativity and healthcare innovation. But it failed at that time to create a true community of changemakers and a lasting impact.

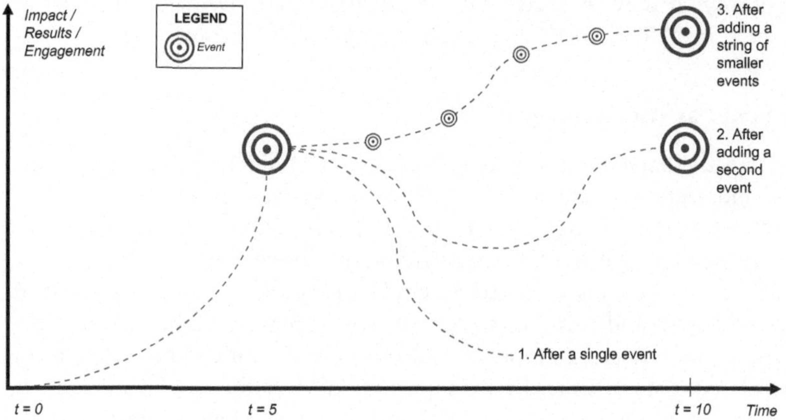

Figure 5.3 Illustration of the impact of holding events on a regular basis.

Quite desperate with the situation in Montréal despite such a successful first event, new leaders rapidly joined the early co-founders. Their intuition at that time was to organise smaller but regular events to bring people together, create, feed, and grow a community. Their intuition proved to be very insightful.

Hacking Health Cafés were born. They became a core element of the methodology used by Hacking Health chapters worldwide. Hacking Health Cafés were described in the previous section. Regularly held throughout the year, they feed the momentum of the Hacking Health community in the cities where it has a foothold. They are particularly instrumental in the preparation of hackathons. They attract and connect individuals who get acquainted with each other, their contexts, needs, and experience, and with the innovation projects led by others. They are a source of knowledge, connection, and inspiration. And they are a source of motivation to actually engage in the innovation process. Most importantly, they feed the community and help it maintain its rhythm and momentum, just like metronomes.

5.7.2 Metronomes to manage risks and go further

Using events as a metronome either gets you to the goal faster or gets you further in the same amount of time, or both. Of course, punctuating their path with several stages forces innovators to break down their ultimate goal into a series of smaller deliverables, making every step more manageable. But there is more than that. Firstly, sequencing a series of missions enables innovators to crystalise accomplishments along the way and build on them as steppingstones for their next steps. Secondly, working in smaller increments enables innovators to better take measures and adjust. Contexts evolve, learnings emerge. Small events along the way allow communities to regroup and share the experience acquired while being in action. Should progress not be as expected, small events allow one to analyse the situation and understand the root causes of their predicament. They become moments to reflect on the challenges they faced, on situations that were not initially anticipated but that became apparent on the ground. Innovators can then create new and better plans, and restart on a better path. Such a mechanism forcing innovators to stop, "fail fast", and restart can be a blessing to help them get ahead on their complex parkour.

Moreover, repeating the "J" shaped pattern of last-minute productivity bursts multiplies the overall progress achieved. Experience and learnings happen faster, enabling us to compound the progress achieved from one event to the other. All in all, more progress is achieved faster. The original end goal could be met over a smaller period. Often, innovation teams, galvanised by their increasing momentum, will choose to beat the target, and achieve more. The final event will be the crowning of their success.

Series of touch-base events are often used by accelerators and accelerator programmes to drive momentum and generate faster and better results for the participating start-ups. The Hacking Health accelerator bootcamps in Canada, as well as similar acceleration programmes such as the Acceler-Health initiatives and the Health Factory programme spearheaded by European Hacking Health chapters, are all good examples of this. The "HH Design challenges" and "cooperathon" events mentioned above are other examples from the Hacking Health case study where strings of smaller events are programmed to create a rhythm and accelerate momentum. In every case, participants work in one to two week-long bursts to achieve a particular set of goals, to maximise their progress before the next get together. During these regular touch-base events, they will report their progress, receive feedback, suggestions, and advice from experts and organisers. In between these events, they can rapidly evolve their design, reach out to potential users, and collect feedback from them or other relevant stakeholders. The most entrepreneurial will recruit new talents or even develop partnerships with strategic players in their field. Participants achieve all of this and more in a much shorter amount of time then they would otherwise, given the impeding deadline created by the next touch-base event. Their metronome effect creates a tempo.

5.8 Events as markers

First, events are opportunities for pulling together the knowledge of a community, the know-how and "know-who" of a particular field. In the Hacking Health case, Hacking Health events are often such "markers of knowledge": opportunities to learn the state of the art in terms of medical practices, contemporary technologies, and so on. Hackathons provide opportunities to learn about new solutions and the skills to implement them. They are markers of available state-of-the-art technologies for participants, and the state-of-the-art changes over time. For example, when the Hacking Health hackathons in 2014 were all about "mobile apps", no one was using artificial intelligence (AI) in their projects to solve health challenges, while starting in 2017, AI became the hallmark or a feature of the majority of projects.

Secondly, and maybe most importantly, events put a stake in the ground for practitioners, markers showing how far they went and what they could achieve. They are markers to use as milestones along a journey, the yardstick by which to measure progress, the steppingstone for the road ahead. Open lab managers should take note. Because they help practitioners share their experience, events become a snapshot picturing how far the community could go. Events provide leaders and decisionmakers with the trigger for pulling together the knowledge, know-how, and "know-who" of a particular field. By asking experts to showcase sectoral knowledge, events help crystallise the state of the art in a particular domain at a particular point in time. They seal it in proceedings and records of all sorts. There again, they become the markers – the snapshots of communities' membership, technology, capabilities, and knowledge at a moment in time. This is typically the case in industrial congresses, scientific conferences and symposiums, board meetings, shareholders' assemblies, town hall meetings and so on. In the Hacking Health case study, the HH Cafés play a bit of that "marker" role, but some Hacking Health organisers decided to go further to fully deliver a "state-of-the-union" effect. Some merged the hackathon format with a TED-Talk like event where various experts would present the state of the art in several relevant domains. Other Hacking Health organisers added complete training programmes to their hackathons to provide participants with the opportunity to get up to date on topics relevant to their innovation quest. Other HH hackathon organisers incorporated an exhibit component, a kind of fair where past participants could exhibit their projects and the progress they achieved. In all cases, these augmented Hacking Health events clearly had the described "marker" effect.

For open lab managers and innovation leaders, like Hacking Health organisers, such markers are more than simple memento boxes of knowledge. They serve as milestones along their journey – the steps by which to measure progress. Knowing where they are and how far they went is invaluable to decide the next steps and plans of action, the ways to go forward, and the route to take. Events are markers for the road behind, and the road ahead.

5.9 Conclusion

We opened this chapter by referring to the overemphasis on the physical dynamics of innovation ecosystems. Physical places indeed play an important role in creative processes. They are the physical implementation of open labs and their most visible component. When properly designed, they can become canvases for creative processes, for innovators to meet, get inspired, and work together, for serendipity to happen. But places without events are like hardware without software: they provide infrastructure and physical spaces, even inspiration and stimulation, but they fail to deliver all the required processes for innovation ecosystems and communities to thrive.

In this chapter, we discussed how events can address these shortcomings. Using Hacking Health as a case study, we first described how events help engage key stakeholders across domains, organisations, and sectors, to bolster creativity, generate new projects, and create momentum. We saw how Hacking Health leaders first use events to attract and connect the right actors, to instil discipline, and drive progress. The Hacking Health case study also helped us expose how events help create and nurture temporal dynamics to bolster innovation communities. We examined how events help their leaders grow and maintain momentum over time within their local community, by programming regularity and repetition in their get togethers. We finally noted that, at Hacking Health or elsewhere, events are used to train, teach, create, share, and crystalise knowledge at moments in times. These dynamics go beyond the temporality elements of innovation ecosystems but also touch their epistemological foundations given how they feed, maintain, and grow the tacit and explicit knowledge of innovation communities.

We have therefore shown that events play five essential functions in effectively fostering communities of innovation and temporally orchestrating innovation ecosystems, from helping to engage key stakeholders to bolstering their creativity, discipline, and momentum, to holding communities together and regularly crystalising the state of the art of their practices. By focusing on such functions of events, we hoped to reveal how they impact organisations and their ecosystems' capacity to innovate more efficiently. We hope this will better equip open lab leaders to achieve greater impact in the pursuit of their mission. A next step in our studies would explore the skills, competences, and know-how they would need to programme, manage, and facilitate events in a way to maximise their results and best achieve the desired impact. The Hacking Health case study could, there again, be a rich source of perspectives on this front.

All in all, the perspectives developed in this chapter overall, and on the functions of events in particular, draw attention to the importance of creating synchronicity. They highlight that temporality should always be taken into account, and not only be considered from a linear point of view. They finally help understand how time can effectively be orchestrated in an innovation ecosystem by organising sets of appropriately programmed gatherings. Overall, this increases our collective understanding on the functioning of innovation ecosystems, for scholars as well as practitioners.

References

Aarikka-Stenroos, L., and Ritala, P. (2017). "Network management in the era of ecosystems: Systematic review and management framework." *Industrial Marketing Management*, 67: 23–36.

Adner, R. (2006). "Match your innovation strategy to your innovation ecosystem." *Harvard Business Review*, 84(4): 98.

Adner, R. (2017). "Ecosystem as structure: An actionable construct for strategy." *Journal of Management*, 43(1): 39–58.

Adner, R., and Kapoor, R. (2010). "Value creation in innovation ecosystems: How the structure of technological interdependence affects firm performance in new technology generations." *Strategic Management Journal*, 31(3): 306–333.

Almirall, E., Lee, M., and Majchrzak, A. (2014). "Open innovation requires integrated competitioncommunity ecosystems: Lessons learned from civic open innovation." *Business Horizons*, 57(3): 391–400.

Autio, E., and Thomas, L. (2014). "Innovation ecosystems." In: Dodgson, M., Gann, D. M. and Phillips, N. (eds.), *The Oxford handbook of innovation management*, pp. 204–288. Oxford: Oxford University Press.

Boudreau, K. J., Lacetera, N., and Lakhani, K. R. (2011). "Incentives and problem uncertainty in innovation contests: An empirical analysis." *Management Science*, 57(5): 843–863.

Briscoe, G. (2014). "Digital innovation: The hackathon phenomenon." *Computer Science*. https://www.semanticscholar.org/paper/Digital-Innovation%3A-The-Hackathon-Phenomenon-Briscoe/cb8e44ec1bcd6062e5fccafb6837030be334731d.

Bullinger, A. C., Neyer, A. K., Rass, M., and Moeslein, K. M. (2010). "Community-based innovation contests: Where competition meets cooperation." *Creativity and Innovation Management*, 19(3): 290–303.

Citroni, S. (2015). "Civic events in a dynamic local field: The role of participation for social innovation." *Industry and Innovation*, 22(3): 193–208.

Cohendet, P., Grandadam, D., Simon, L., and Capdevila, I. (2014). "Epistemic communities, localization and the dynamics of knowledge creation." *Journal of Economic Geography*, 14(5): 929–954.

Corley, K. G., and Gioia, D. A. (2004). "Identity ambiguity and change in the wake of a corporate spinoff." *Administrative Science Quarterly*, 49(2): 173–208.

Dedehayir, O., Mäkinen, S. J., and Ortt, J. R. (2018). "Roles during innovation ecosystem genesis: A literature review." *Technological Forecasting and Social Change*, 136: 18–29.

Eisenhardt, K. M. (1989). "Building theories from case study research." *Academy of Management Review*, 14(4): 532–550.

Fine, G. A. (1990). "Organizational time: Temporal demands and the experience of work in restaurant kitchens." *Social Forces*, 69(1): 95–114.

Granqvist, N., and Gustafsson, R. (2016). "Temporal institutional work." *Academy of Management Journal*, 59(3): 1009–1035.

Granstrand, O., and Holgersson, M. (2020). "Innovation ecosystems: A conceptual review and a new definition." *Technovation*, 90: 102098.

Johnson, P., and Robinson, P. (2014). "Civic hackathons: Innovation, procurement, or civic engagement?" *Review of Policy Research*, 31(4): 349–357.

Jolly, S., and Raven, R. P. J. M. (2016). "Field configuring events shaping sustainability transitions? The case of solar PV in India." *Technological Forecasting and Social Change*, 103: 324–333.

HH Website. (2020). www.hacking-health.org.

Kaplan, S., and Orlikowski, W. J. (2013). "Temporal work in strategy making." *Organization Science*, 24(4): 965–995.

Lampel, J., and Meyer, A. D. (2008). "Field-configuring events as structuring mechanisms: How conferences, ceremonies, and trade shows constitute new technologies, industries, and markets." *Journal of Management Studies*, 45(6): 1025–1035.

Lampel, J., Jha, P. P., and Bhalla, A. (2012). "Test-driving the future: How design competitions are changing innovation." *Academy of Management Perspectives*, 26(2): 71–85.

Mérindol, V., and Versailles, D (2017). "Développer des capacités hautement créatives dans les entreprises: le cas des laboratoires d'innovation ouverte." *Management International/ International Management/Gestiòn Internacional*, 22(1): 58–72.

Moore, J. F. (1993). "Predators and prey: A new ecology of competition." *Harvard Business Review*, 71(3): 75–86.

Murray, F., Stern, S., Campbell, G., and MacCormack, A. (2012). "Grand innovation prizes: A theoretical, normative, and empirical evaluation." *Research Policy*, 41(10): 1779–1792.

Nambisan, S., and Sawhney, M. (2011). "Orchestration processes in network-centric innovation: Evidence from the field." *Academy of Management Perspectives*, 25(3): 40–57.

Phillips, M. A., and Ritala, P. (2019). "A complex adaptive systems agenda for ecosystem research methodology." *Technological Forecasting and Social Change*, 148: 119739.

Puckrein, G. (2016). "Nothing about me without me: Patient research exchange." https://www.patientresearchexchange.org/stories/detail/nothing-about-me-without-me.

Walrave, B., Talmar, M., Podoynitsyna, K. S., Romme, A. G. L., and Verbong, G. P. (2018). "A multilevel perspective on innovation ecosystems for path-breaking innovation." *Technological Forecasting and Social Change*, 136: 103–113.

Yin, R. K. (2009). *Case study research, Design and methods*. Los Angeles and London: Sage.

6 Communitech in Waterloo, Canada

How open lab organisations can drive a successful entrepreneurial ecosystem

Luc Sirois, Octave Niamié, and Patrick Cohendet

The Kitchener-Waterloo region in Canada is one of the most dynamic entrepreneurial ecosystems in the world. In its 2020 report, the scoring organisation, Canadian Tech Talent (CBRE, 2020), presented it as the ecosystem with the most dynamic technological talent market in Canada with one of the highest rates of tech job growth in the country. According to the latest Startup Genome report (Startup Genome LLC, 2020), the Toronto-Waterloo tech entrepreneurial ecosystem is the 18th most successful worldwide. This chapter analyses some of the factors that contributed to the formation and ongoing momentum of this successful entrepreneurial ecosystem, focusing in on the contribution of Communitech, an "open lab" type of organisation dedicated to "helping businesses start, grow, and succeed" in this region.

An entrepreneurial ecosystem can be defined as a set of actors and factors that are interdependent and coordinated in such a way as to foster productive entrepreneurship (Stam, 2015; Stam and Welter, 2020). It can also be defined as "a set of interconnected entrepreneurial actors (both potential and existing), entrepreneurial organisations (e.g., firms, venture capitalists, business angels, banks), institutions (universities, public sector agencies, financial bodies), and entrepreneurial processes (e.g., the business birth rate, numbers of high-growth firms, levels of "blockbuster entrepreneurship", number of serial entrepreneurs, degree of sell-out mentality within firms, and levels of entrepreneurial ambition) which formally and informally coalesce to connect, mediate, and govern the performance within the local entrepreneurial environment".

Indeed, Communitech is considered to be one of the architects of the success of Waterloo's technological entrepreneurial ecosystem. Founded in the 1990s by a group of tech entrepreneurs aiming at pooling effort and knowledge to help each other succeed, the association started in the form of informal interactions, before becoming incorporated in 1997 with 43 members. It later built its own physical place, the Communitech hub, and now bolsters more than 1,600 member companies comprising start-ups but also corporations and scale-ups in the process of becoming global leaders.

Researchers have closely analysed the role of incubators or accelerators and "open labs" organisations like Communitech in building a successful entrepreneurial ecosystem (Drori and Wright, 2018; Goswami et al., 2018;

DOI: 10.4324/9781003125587-9

Roundy, 2021). These studies reveal that they generally play the role of intermediary between elements of the ecosystem (Drori and Wright, 2018). Business incubators/accelerators/open labs are also considered elements of the social attribute of an entrepreneurial ecosystem, providing support in the forms of infrastructure, training, or networking opportunities.

These studies, however, provide limited insight of the role of open lab organisations in the dynamics of the entrepreneurial process. They tend to consider these hubs as a given. Most importantly, they ignore how successful ones such as Communitech make a real difference for entrepreneurial ecosystems by linking their contributions – support services and space – to strong community values.

In this chapter we aim to bridge this gap in the literature by investigating the role of a value-driven entrepreneurial community and its open lab organisation in raising a high-performing entrepreneurial ecosystem by using the case of Communitech. To this end, we carried out documentary research and a series of semi-structured interviews with a range of people directly or indirectly involved in the mission of Communitech. We interviewed political actors, coaches, management staff, and entrepreneurs, 25 people in total.

Our analysis and this chapter are structured as follows. We first present the role of key actors, local community values, and approaches used in fostering the emergence and evolution of this successful entrepreneurial community. Secondly, we expose how Communitech purposely aimed at orchestrating, and adapted to, the evolution of the local entrepreneurial ecosystem. Thirdly, we explore in more depth the anchoring of Communitech and its community in a physical place, the Communitech "Hub" at the Lang Tannery. We conclude by putting the evolution of Communitech into perspective.

6.1 Fostering the emergence of an entrepreneurial ecosystem: key actors and distinctive approaches

Among elements that contributed to the success of this entrepreneurial ecosystem are the regional culture of collaboration as a fertile ground, the presence of an entrepreneurial university and an active community of entrepreneurs. We discussed the roles played by these key actors, how they transposed the regional culture into Communitech's approach and inspiring vision, and some of the best practices and approaches that were used to achieve success.

6.1.1 Regional culture as a fertile ground

Our interviews revealed that the value of collaboration that brings these passionate tech entrepreneurs together is embedded in the Kitchener-Waterloo region's high levels of collaboration and knowledge sharing. The attributes associated with the regional culture are collaboration, mutual support, ambition, and innovativeness. These attributes are repeated on various platforms and by various community leaders. As an illustration, the following inscription is written on Communitech's website (2021a): "Waterloo Region's founders had a strong tradition of coming together to get big jobs done quickly (and well)". This way of articulating the regional values not only has the effect of making them explicit and communicated, but it also promotes and reinforces them.

This desire to collaborate to get things done is well anchored in the DNA of the people of the region who prize cooperation at the heart of their culture. This region is recognised to be resilient, with an ever-present capacity to reinvent itself. From being famous for its manufacturing industry, then for its insurance industry, this region is now known as the most vibrant technology entrepreneurial ecosystem in Canada.

These values have been reported to be rooted in the region's German and Mennonite heritage. However, Spiegel and Bathelt (2019) reported that the Germans in the Kitchener-Waterloo region do not make strong use of their social capital for professional purposes. They wrote that the idea of a "'German culture' of cooperation, innovation, and trust between firms and industries as an explanation for the region's success as a technology economy is simply a discourse … that has helped create the strong sense of community within the region upon which a well-recognized entrepreneurial ecosystem has developed" (Spigel and Bathelt, 2019: 278–279).

If the link to the Mennonite heritage of cooperation and trust is thin at best, the proactive communication of these values not only promotes them but contributes in reinforcing the ecosystem. Constantly calling on these values becomes a means to link together different actors, actions, and other elements of the region's entrepreneurial ecosystem.

6.1.2 The University of Waterloo

An efficient ecosystem relies on key players, including universities. The role of the University of Waterloo (UW), "a university built on, and for, industry collaborations and entrepreneurship", in the emergence of the local community of technological entrepreneurs is underlined by several sources (Bathelt et al., 2011; Bramwell, 2008).

It is reported that in a time when mathematics was considered an abstract science, the head of the mathematics department of the UW had a different view. He saw that mathematics was going to be part of our daily life, as this article highlights:

> The University of Waterloo began with engineering, mathematics, and science. […] The head of the math department, Ralph Stanton, had the vision to see that his field would become increasingly integral to modern life. He had written a textbook on numerical analysis, a branch of mathematics that is closely aligned to computing. In 1960, the university established its Computing Centre, and suddenly math had a practical application. The department grew so quickly it was expanded in 1967 into a separate faculty, the first in North America …
>
> (Gillmor, 2012)

This university has a reputation of putting expensive IBM machines in the hands of undergraduates to help them learn and develop skills, which was contrary to mainstream practices of protecting expensive equipment. UW is also considered as one of the strongest proponents, some would say the inventor, of the Co-op Programme for students. Now scattered in universities nationwide, this approach integrates paid internships in companies for students to complete over the course of their university degree. This collaboration initiated by UW allowed the academic world to better understand the real problems of the industry, to develop appropriate solutions, and to better prepare students for the real world. With this approach and many more, UW produces highly trained talents that support innovative companies, which in return attract talented tech workers from around the world, creating a virtuous cycle that strengthens the Kitchener-Waterloo entrepreneurial ecosystem. UW researchers are known to be entrepreneurial thinkers and industry partners. It supports the local industry by making research and development resources available to it. UW has a unique Intellectual Property (IP) Rights principle anchored in the local culture of sharing and collaboration called Policy #73 or "creator owned", which grants ownership to the inventor. This policy and the university's entrepreneurial culture have positioned Waterloo as a national leader in the transfer of technology to the private sector (UW, 2021).

6.1.3 The community of tech entrepreneurs

Successful high-growth tech companies acted as initial anchors. The virtuous synergy between industry and UW, as well as the availability of talents, has led to the emergence of successful tech companies, in particular Research in Motion (RIM) in 1984 or OpenText in 1991. These successful companies (and many more) created thousands of well-paid jobs in the community (even after the demise of some of them like RIM) and fostered the emergence of several other tech companies. To date, more than 700 companies belonging to researchers, students, or alumni are active in the region (UW, 2021).

Communitech initially started from an initiative launched by entrepreneurs heading such promising tech companies. They were passionate entrepreneurs and provided the initial impulse. They decided to come together and pool their efforts and knowledge to help each other succeed and created an organisation 1997 to support the entire "Community of Tech" with 43 founding members. They wanted to address shared issues of a growing number of peers in the face of local resources issues and other pressures, such as increasing global competition impacting them locally. While they focused first on the desire to collaborate and help each other, they hoped to convince key decisionmakers to address some of the systemic issues they were facing. They also aspired to drive the emergence of other strong local tech companies like theirs. In the process, they formed and developed a strong entrepreneurial community in the Waterloo area, held together by Communitech. This was later made explicit in the ways Communitech communicated about itself, as illustrated by this quote from its CEO:

> We're a community of collective insights, experience. and wisdom. Tackling challenges together and having each other's backs allows us to have more influence and make the most of our combined networks. We're fast. We're connected. We're trusted.
>
> Klugman, CEO (Communitech, 2016)

6.1.4 A bigger-than-oneself manifesto

From its origin through to today, as just an organisation and later as a place, Communitech always explicitly anchored its mission in the collaborative values of the Waterloo area. It describes itself as a place and a group with "a pact" to "help companies". It constantly and systematically emphasises its value-driven approach, as illustrated by this excerpt from its website:

> Communitech helps tech companies start, grow and succeed. That's our mission, our mantra, our reason for being. Everything we do ties back to collaboration and helping values that run deep in our organization.
>
> (Communitech, 2021)

Moreover, Communitech has adopted a vision that transcends the direct and immediate interests of entrepreneurs and extended to the global leadership of the region. This vision, reported on its various communication platforms, is an inspiring "bigger-than-oneself" manifesto:

> Building one of the most successful start-up ecosystems in the world.
>
> (Communitech, 2021)

It was more than the vision of Communitech, it became the rally cry of an entire entrepreneurial ecosystem.

6.1.5 An associative model

Unlike many recognised innovation places where tenants stay for a limited period (Cohen, 2013), Communitech has developed an associative model that allows entrepreneurs to be members of their ecosystem without any space or time limitation. Moreover, the associative model of Communitech makes it possible to unite actors more concretely around a common vision. Through inspiring rhetoric around a common vision and mission for the group, it contributes to convey the deep collaborative values of the community to members. It helps maintain or strengthen the bonds between members of the community and to develop a sense of belonging, each considering themselves as a "member of a unique club". It incidentally increases everyone's sense of responsibility for the success of Communitech's Hub, its "clubhouse". All in all, the associative model generates a positive dynamic among companies and entrepreneurs unified by common objectives, memberships, communications, platforms, and events – which would probably not have happened otherwise. All of these elements enhance the attractiveness and strength of the community. Finally, the associative model also contributes to the long-term financial viability of Communitech, as it provides a relatively stable recurring income stream and a larger addressable market compared to the traditional incubation model whose income comes mainly from logistics services offered to members (rental infrastructure). To date, Communitech has more the 1600 members including 800 start-ups, 140 scale-ups, 100 large corporations and more than 500 service providers, governments, academia, and not-for-profit organisations (Communitech, 2021). We consider the membership model one of the keys to the Communitech's success.

6.1.6 Determined leadership

Another factor behind Communitech's success is the leadership of its CEO, Iain Klugman, for over 17 years, until very recently (May 2021). Contrary to the original founders, of which he was not a member, Klugman was not a tech entrepreneur before joining the organisation but rather an executive in a leading media organisation. He had also previously worked in the fields of tourism and telecommunication. His skills and involvement made a difference in the development of Communitech and the community, according to numerous colleagues and analysts, as noted in the press and private interviews.

> Under Klugman's leadership, Communitech grew from a little-known local industry association into a globally recognized innovation hub, providing facilities, programming, and advocacy to a dynamic and growing community of 1,400 technology companies in Waterloo Region.
>
> (Communitech, 2021)

Upon taking the reins of Communitech, he quickly became a vocal advocate for entrepreneurial and regional development, constantly using his influence to create a virtuous circle of community mobilisation in the region, and later in the province. Despite this high profile, Communitech's success was not the heroic exploit of a strong individual alone. It was rather a collective success fuelled by a leader's distinctive efforts to bring people together and, most importantly, to constantly communicate the community's values of collaboration, mutual support, and ambition. He and his team repeatedly demonstrated a unique capability to rally the support of community leaders, to mobilise stakeholders of the Kitchener-Waterloo entrepreneurial ecosystem, and to foster a common vision. They always did so with an inspiring rhetoric and clear messages, a likely contribution of Klugman's past in the communications and media sectors.

His determination earned him a reputation for being particularly intense in his approach to enlist key people in Communitech's mission. For example, Klugman makes personal contact with key decisionmakers and surrounds himself with top community leaders when necessary to make his effort more impactful.

6.2. Aiming to orchestrate the entrepreneurial ecosystem … and to evolve with it

In this section, we discuss how Communitech dedicated itself to be a catalyst and orchestrator for the Kitchener-Waterloo entrepreneurial ecosystem and how its mission evolved over time in response to the maturing of the entrepreneurial community, which is typical of best practice open labs which manage and adapt a portfolio of services based on the evolving needs of their stakeholders.

Contrary to what can be found in many other cities, the development of the tech ecosystem in the Kitchener-Waterloo region is not dispersed over a very large number of organisations and networks. While not alone, Communitech played a central role of catalyst in this ecosystem. Through its leaders, it aimed at concentrating the local forces, at interlocking key strategic players. It purposely tried to orchestrate the development of the entrepreneurial ecosystem. At one point in time, it even defined itself as a form of economic development agency.

> [*Communitech*] operates as a hybrid economic development agency, marketing board and business support network.
>
> (Dingman, 2015)

We believe its collaborative and inclusive mindset, its value-driven approach earned it the "moral right" and credibility to play such a role. Its inspiring vision for the future of the region, its mission-driven actions (as explained in the previous section) were powerful enablers in this sense. As explained above, Communitech's vision – "Building one of the most successful start-up ecosystems in the world" (Communitech, 2021) – is inspiring. With such a statement, Communitech's leaders clearly articulate that their goal is not to simply provide a place or programming for businesses, but to orchestrate the ecosystem and bring it to new heights. This vision creates a manifesto "bigger than oneself" and positions Communitech as an organisation on a mission. It gives the leaders the ability to unite dispersed stakeholders, and naturally positions them to attract more and more support and resources along the way.

In addition, their ability to adjust their mission and focus over time allowed them to remain relevant, to adapt to changing opportunities and to evolve with a maturing community.

From start-ups …

The mission of Communitech is to support technology companies and is clearly and concisely expressed as follows: "Communitech helps tech companies start, grow and succeed", by providing them with the space needed and a mentoring programme. Thus, from an association originally aiming at bringing together successful entrepreneurs wishing to collaborate to better compete internationally, Communitech became institutionalised and carried on a mission of economic development that consists of helping businesses start up, grow, and succeed. In collaboration with other players from the ecosystem such as the UW, they contributed to the emergence of a dense network of start-ups.

… to corporations …

Whereas in the beginning the focus was more on helping start-ups, in the 2010s Communitech broadened its field of action and announced its interest in helping large corporations in the region to innovate. Based on the assumption that large corporations lacked the agility to innovate, the leaders of Communitech decided to put them in contact with innovative start-ups to help them in their innovation processes, while allowing these young companies to "have access to these corporations as their first market to test their business models" as stated below.

> Big companies get to tap into Waterloo's legendary tech talent, develop ideas far more quickly and tune their workplace cultures to the new economy, while our start-ups get access to the buying power that big brands represent.
>
> (Klugman, 2015)

Several large corporations, ranging from retailers to banks, set up innovation labs at Communitech. In 2013, the retailer Canadian Tire opened the first corporate innovation lab at Communitech. Shortly after, several major companies followed suit, with TD Bank Group, Manulife, Canon, Deloitte, Thomson Reuters, and others opening their own innovation labs there. The objective is to give them access to a space where they can safely experiment new things but also contribute to the success of the region which has collaboration at the heart of its culture, as mentioned in the following quote:

> Communitech, situated in the heart of Canada's most dense start-up community and right next door to the country's corporate and financial capital, is ideally positioned to help these large firms – not only by providing offsite facilities where they can try new approaches, but by plugging them into Waterloo Region's greatest asset: an entrepreneurial ecosystem teeming with world-class talent, built on a strong local culture of collaboration.
>
> (Klugman, 2015)

Large companies saw their coming to Communitech as an opportunity to innovate through interactions with innovative start-ups. Communitech sees this collaboration between corporation and start-ups as a market opportunity for entrepreneurs and welcomes both in its hub. This reinforces Communitech's status as the middleground where the upperground made of big corporations interacts with the underground of start-ups to better innovate.

… to scale-ups

Over time, we could then observe a migration of Communitech's attention to scale-ups. In 2017, in an article published on the Communitech website, the organisation declared that its main objective is to help companies to succeed and grow, thus omitting the word "start" which, along with the other two, used to convey the mission of this organisation. We present here this excerpt from a text that puts Communitech's evolution into perspective and reflects the migration of its interest to growing companies.

> In just 20 years, Communitech has followed a similarly impressive evolutionary arc – from its humble beginnings in a corner of Taaz Corporation's head office to its current incarnation in over 110,000 square feet of the iconic Lang Tannery, the organization set out to develop one of the most successful tech ecosystems on the planet by helping companies succeed and grow.
>
> (Communitech, 2017)

This change was also felt in the interviews we had between February and March 2021 with workers and people close to Communitech. We learned through these interviews that Communitech is more oriented towards growing companies in order to support them to become world leaders with a very strong impact on the development of the region. Klugman even shared a co-development activity with us that he himself runs periodically, enabling scale-ups' leaders to collaborate to better address their common challenges.

In 2019, Klugman told the press that Waterloo was no longer going to be known only as the cradle of start-up technology companies but a place where technology companies capable of competing on an international level emerge:

> This is the year that Waterloo Region will shift from being a start-up hub to being the home of several new globally competitive tech companies. I predict we see at least one new company reach $100M in revenue and see at least one really significant exit. … We'll also see the top 25 scaling companies double their hiring (from 1,000 new employees hired in 2018 to at least 2,000 in 2019).
>
> (CBC, 2019)

In summary, Communitech has become a place at the heart of the Kitchener-Waterloo entrepreneurial ecosystem by adopting an open lab logic. It is a place where companies of all sizes converge to access knowledge, know-how, creativity, and resources available in the community to innovate and conquer markets (Mérindol and Versailles, 2016; Schmidt and Brinks, 2017). The UW has even set up its Velocity incubator there as we will see below. Even the government sees it as a suitable platform to better reach the entrepreneurial community of the region. It conducted, through Communitech, several support initiatives including the recent Digital Main Street (DMS) set-up to help small businesses go digital (Communitech, 2021b).

6.3 Embedding their local "community of tech" values in a place

Given such a strong value- and mission-driven approach, we analyse in this section how the values of Communitech drove the creation of a key element of its value proposition, namely a physical innovation "hub", to maximise its overall impact. We also examine the importance of "programming" in achieving such impact and in support of the success of the place.

Scholars have been trying to distil the key factors of success of thriving entrepreneurial ecosystems for decades – trying to identify the ingredients that most positively impact the socioeconomic development of a region. Among such ingredients, "location" had initially been "taken for granted, its influence underappreciated or … controlled away" when considering innovation ecosystems (Welter, 2011).

Some scholars started to explore the dimension of "place" with a desire to understand "how the cultural, social, political, and economic structures […] associated with a place" impact the entrepreneurial outcome (Spigel and Stam, 2016). However, their notion of "location" or even "hub" are often of a regional scale, typically extending around cities or metropolitan areas.

At the same time, "places" at the scale of a single building or workplace have emerged everywhere and are the hallmark of all government or corporate-led entrepreneurial and innovation development strategies (Moultrie et al., 2007; Oksanen and Ståhle, 2013). Coworking spaces, start-up accelerators, makers labs, research institutes, innovation hubs, research parks, and the like have clearly become tangible demonstrations of local efforts to showcase the vitality of their entrepreneurial ecosystems. Most importantly, such "innovation places" were identified as one of the important elements of the mix, driving creativity within and between communities present in innovation ecosystems (Grandadam et al., 2013).

6.3.1 Communitech adds a "place" to its open lab value proposition

Following similar impetus, Communitech at one point in its history decided to build and to move into its own innovation space. Incidentally, this was as part of a city's rejuvenation effort that they helped spark and in a completely refurbished industrial factory, the Lang Tannery building, at the heart of an otherwise deserted neighbourhood, powerfully reflecting the will and the capacity of this region to reinvent itself. We believe this move to a physical place, and the importance of "place" in Communitech's value proposition, are key factors in the success of the organisation and entrepreneurial community.

The innovation place called the "Communitech Hub" opened a decade after the incorporation of the organisation but rapidly became a powerful element of its value proposition, a powerful means to help achieve its mission. The "Hub" is presented as the centre of gravity of the ecosystem, a "clubhouse" for the members and visitors. It helped Communitech to strengthen its leadership in the entrepreneurial ecosystem of Kitchener-Waterloo and beyond. Each year, the Hub attracts tens of thousands of local and international visitors, thousands of event attendees and hosts nearly 200 tenant companies on site. The Communitech Hub has become a strong part of the "Waterloo brand" and is considered a flagship model among innovation spaces in Canada. Below we expose some elements behind its success.

Scholars explain that the value proposition of innovation places is made of three essential components: logistics, support, and mediation (Bergek and Norrman, 2008; Bruneel et al., 2012; Peters et al., 2004). Communitech illustrates these three dimensions. Logistics refers to the infrastructure made available to entrepreneurs. These are offices, meeting rooms, telecommunications infrastructure, or laboratories in some cases (Peters et al., 2004), and administrative services (Bergek and Norrman, 2008). These are tangible services that help entrepreneurs reduce their operating costs (Grimaldi and Grandi, 2005). The support services include training offered to entrepreneurs in the form of workshops or seminars (Peters et al., 2004), coaching, mentoring, and business advice (Bruneel et al., 2012; Pauwels et al., 2016). The supports are intangible value-added services that aim to enhance the skills of entrepreneurs (Grimaldi and Grandi, 2005). The mediation refers to the way in which innovation spaces like business incubators or accelerators put entrepreneurs in contact with external actors (Bergek and Norrman, 2008) such as consultants, clients, suppliers, or investors (Bergek and Norrman, 2008; Bruneel et al., 2012; Pauwels et al., 2016; Peters et al., 2004). The purpose of the mediation is to facilitate the entrepreneurs' access to new resources.

In the case of Communitech, creating a physical place had several additional benefits. It helped Communitech provide structure to its entrepreneurial community by materialising it, making it become "more real". Officials and VIPs could show and visit such a hallmark of their community. It became a location that members could go to or gravitate around, providing them with a common focus and connection. Moreover, it provided the Communitech open lab organisation with a clear and established revenue model.

We discuss below how the Communitech Hub place and programming implemented these elements, and some of the approaches and factors that made them successful.

6.3.2 Co-location of key actors

At the Communitech Hub, part of the value proposition is the co-location of several key players of the entrepreneurial ecosystem, to provide access to expertise, capital, services, and connections. The co-location of strategic actors happens at two levels: co-location of key ecosystems actors within the Tannery building, and co-location of corporations with start-ups within the Communitech Hub itself. Moreover, the co-location of such players is often implemented with seamless boundaries between spaces, creating a greater sense of integration and of critical mass for the ensemble.

6.3.2.1 Co-locating key ecosystems actors within the Tannery building

The co-location within the Tannery building of the Communitech Hub with strategic actors – UW's Velocity incubator, Google, D2L – played a strategic role in the success of this innovation place. It creates and pulls together of a broader ensemble and bolsters the reputation of the place. Most importantly, it better supports the overall mission and is yet another way to integrate the values of collaboration and mutual support.

For the Lang Tannery building owner, this co-location strategy contributed to a more sustainable business model, stemming from diverse rental rates, with financially solid tenants covering more of the overall real estate cost of the building than the Communitech Hub itself. It provided more services and reasons to come to the space for key stakeholders. It created a broader set of powerful organisations – including the University of Waterloo and Google – to root for the place. At the end, it helped reach a critical mass of start-ups, people, and resources.

In particular, the co-location with the Velocity incubator sustains the community's mission. Velocity is considered one of the most impactful free university-linked incubators in Canada. It remotely engages with approximately 1,500 students each term on campus and their companies have received nearly C$1 billion worth of investment. With its expansion within the Tannery completed in 2016 it became the "largest free start-up incubator in the world" (UWaterloo, 2016). As a high-performance breeding ground for tech start-ups (and the origin of some of the region's most famous ones) Velocity directly contributes to the Communitech Hub success. It also allowed the Communitech organisation to focus on ecosystem orchestration and other critical services and programming, as opposed to trying to be an incubator itself.

Finally, while its impact is unclear, one should also notice the onsite location of artisan workplaces. The co-location of such workshops within the building was originally seen as the way to integrate "The Arts and Culture Scene" in the concept as a potential trigger for further creativity, innovation, and out-of-the-box collaborations. However, further research would be necessary to assess whether these objectives were achieved.

6.3.2.2 Co-locating large enterprises onsite and along start-ups

Communitech hosts several corporate "innovation labs", which are creatively designed corporate offices right at the heart of the Tannery building with the specific purpose to help large enterprises innovate in today's context of rapidly changing technologies, markets, and business models. It surrounds these large enterprises with start-ups in a dynamic environment both physically and in terms of programming. The goal of this corporate innovation programme is to "inject start-up DNA" (Communitech, 2021) into large corporations, help them "explore ways to disrupt", and create new opportunities through collaborations and their participation in an ecosystem.

> Communitech's corporate innovation program is focused on helping big brands stay competitive and manage disruption. Dedicated account managers and Communitech's Innovation Playbook help our corporate partners develop behaviours and a timeline of activity to get the most value out of their ecosystem engagements. We help our corporations engage with start-ups, develop new products quickly, foster a culture of innovation, and attract and retain the best talent. In return, they make Waterloo Region stronger by being marquee customers, investors, and partners to local companies, bringing a global perspective and big-brand name recognition to our ecosystem.
>
> (Communitech, 2016)

Such strategies attracted financially strong corporations like TD Canada Trust, Deloitte, GM, LCBO, Manulife, Thomson Reuters, etc., and create value for the Hub in multiple ways. It provides a key element of the Hub's financial sustainability model with profitable income. It contributes to the virtuous circle of credibility, reputation, support, and attractiveness towards other stakeholders. It also furthers the culture of collaboration and mutual support, at the heart of Communitech's approach.

6.3.3 Best practice layout and design

The literature exposes how a mix of a variety (Imai and Ban, 2016; Steelcase, 2015) of zones in a workplace is important to foster creativity: meeting zones, social zones bringing people together to support serendipity and the creation of a community, isolation zones for quiet individual work, relaxation or rest, nomadic zones to support on-the-go individuals and their particular needs, as well as a resident zone providing a permanent space for individuals and teams – all with attractive design elements from aesthetic and even psychological points of view, such as arts and light. But one must go beyond the sheer elements of design and architecture in creating a successful innovation place. Physical attributes must include, for example, the presence of lifestyle services, open access to part of the venue, onsite presence, and accessibility of key actors of the entrepreneurial ecosystem.

Both the Tannery building and the Communitech Hub inside it showcase several such best practices, not only creating a successful innovation place but also a "lifestyle environment" for its users. The key elements, layout, and division of the place underline the Communitech values of collaboration, connection, and mutual support. Several essential ingredients are present.

6.3.3.1 Important layout and design elements

First, the innovation place was created out of an old industrial building, completely refurbished with a dedicated attention to key architectural dimensions. "[The architects] preserved the masonry shell of the Tannery building and reduced the maze of closed-in buildings" (Architectural Record, 2011). Setting up in an abandoned industrial area and refurbishing an old factory is a strong symbol of the capacity of this region to reinvent itself, and to replace a declining manufacturing industry with an emerging tech industry.

A café and "fun space" reins in the middle of the Hub and is a centre of attraction for all onsite workers and visitors, acting as a "watering hole". By design, it becomes a gateway for residents to meet and connect with external actors – mentors, investors, and others – and invite individuals to socialise. The spirit and energy level emanating from this core area truly nourishes the sense of community that Communitech prides itself about.

Besides such an "informal" meeting area, formal meeting spaces in and of themselves are fundamental elements of successful innovation spaces. Communitech features a wide range of meeting room sizes, one of which borders the central café, which allows informal interactions to transition to more formal meetings and vice versa, and to convey more broadly the energy and spirit of the place even in the formal meetings. Small offices are available for start-ups graduating from the open coworking space which require larger spaces with more privacy. These closed-in offices represent a transition for growing companies before they further graduate to renting their own offices outside the Communitech Hub.

A local café and microbrewery pub is located onsite, outside of the Communitech Hub but within the Lang Tannery building. While this appears anecdotal, we perceive such a venue as an important element of a successful innovation place in general and of Communitech in particular. It too acts as a watering hole for the community, contributes to the "lifestyle" and "social" element of the place and enables a greater variety of encounters with visitors.

Other notable elements of the Hub's value proposition are an event venue and speciality areas. The speciality areas – the "science lab" and "hardware workshop" – make tools and equipment available for tenants, enabling hands-on work and supporting other niche needs. The event venue is large, fully equipped, and allows for organising conferences, training, and other formal gatherings, including social and community functions.

6.3.3.2 Seamless integration of spaces

The layout of the building presents an open integration of the different office spaces occupied by the different tenants. Their offices are either closed-in and separated when necessary, such as for Google and other offices, or wall-less and seamlessly co-located when a more fluid delimitation between the two was desirable, such as between the resident corporations among themselves, or between the Velocity incubator and the Communitech Hub.

The absence of walls and traditional closed-in offices makes co-location seamless with an overall integrated feel as if all components, albeit independent, belonged together. The perception of fluid integration of these strategic partners was and is fuelled with a purposeful ambiguity around the Communitech Hub's physical boundaries, as most public communication leads observers to believe the whole Lang Tannery building is the Communitech Hub, as opposed to just a 44,000 square feet space within it. The more impressive size of the ensemble, the presence of brand-named tenants, the combined amount of activity happening under one roof, and the greater number of people onsite, generate a greater buzz and reputation for the place and Communitech. This in turns makes Communitech grab more attention and fascination from key local and national decision makers.

This "seamless integration" approach virtually creates and pulls together a broader "ensemble". The result is more visually and spatially impressive for the visitors and more stimulating for the "residents", such as the large corporations and entrepreneurs, who can see more things going on, more people in action. They can witness more interactions, activities, technologies being built and tested and presentations happening. Such a concept improves the possibility of sharing services between tenants, such as the central bistro and fun space, by making them more visible and accessible. More importantly, it facilitates face-to-face interactions as it uses "the bones of the building" to create a lively area that will attract individuals, enable idea sharing and create a sense of community – all key ingredients of a successful innovation place. It also enables more collisions between the occupants with more open intersections between offices and more possibilities to "bump into" other people. It contributes to build, maintain, and feed the culture of collaboration of Communitech.

6.3.4 The programming

Another key element of Communitech's value proposition is the programming. Programming is an essential component of any open lab's effort to orchestrate the development of entrepreneurial communities, and an essential addition to any innovation "place" strategy. An innovation place without programming would be like a computer hardware without software: just raw potential.

Communitech's programming, quite strategically, comprises a set of activities that connect innovators to education, mentoring, and meaningful collaboration, creating opportunities for introductions and learning, and connections to necessary resources. The programmes themselves are exhaustive and focus on providing support and skill development in strategic areas, for companies at all stages of the economic value creation continuum. They all integrate the values of collaboration and mutual support. Here are some of these elements:

Specialised skills development: "The Communitech Academy is a trusted partner in developing talent to meet the needs of tech and tech-enabled organizations" (Communitech, 2016). It focuses on upgrading employee skills in broad areas relevant for the tech sector and modern workplace, from hard technical skills to new professional skills and best practices in some key business areas.

Talent recruiting and retention: Communitech provides a number of resources, tools, events, and services to help tech companies to attract and develop talents as well as a number of events and activities dedicated to attracting international talents towards the Kitchener-Waterloo tech sector.

Hands on support for start-ups: Communitech creates conditions for start-ups to receive support from their peer groups. The first stage of the support is typically in the form of short sessions with panels. The more deserving start-ups will be assigned a "Lead Advisor" or "Growth Coach" to advise and support on a more ongoing basis.

Support for scale-ups: advisory services, training programmes and networks focused on the particular needs, opportunities, and challenges of companies with rapidly growing revenues. It focuses on three main areas: access to customers, revenue, and talent.

Business support services: Communitech built specialised teams and services targeted at the critical pain points and emerging opportunities of new and fast-growing companies: "Doing more with data", "Getting in front of customers", "Getting in front of money", "Expanding sales in new markets". "Pro Squad" provides onsite access services from business professionals.

As mentioned in the opening, the literature on innovation spaces informs us that these actors design their value positions around three axes: first of all, logistics services, then support, and mediation (Bruneel et al., 2012). However, we know very little about how these dimensions of their value propositions are combined and delivered to meet the demand of the entrepreneurs. In the case of Communitech, we found that the entire value proposition is designed in a manner that reflects the community's cultural values of collaboration, mutual support, ambition, and innovativeness. In particular, the Communitech Hub physical place is designed in such a way as to encourage unexpected encounters between occupants. Its co-location strategy promotes entrepreneurs' access to certain key resources. The support activities, the programming, are designed in such a way as to create the know-how and links between many different actors, from start-ups to corporations, to scale-ups and resource persons.

6.4 Conclusion and perspectives on the road ahead

Founded in 1997 by prominent local tech figures committed to making the Waterloo region a global innovation leader, a small yet influential organisation stood behind many of the region's successes: Communitech. Ten years later, as part of a city's rejuvenation effort that it helped spark, Communitech built and moved into its own innovation space in a completely refurbished industrial factory at the heart of an otherwise deserted neighbourhood, and since then redefined itself as "a place, a posse, a pact" "to help tech companies start, grow and succeed".

Communitech is a unique model of open lab and innovation space in that it emanated from a community of tech entrepreneurs who decided to join forces to support each other, to innovate and better overcome the challenges they face. This association subsequently became institutionalised by taking a form that is often associated with an open lab, but it goes beyond. In ways far beyond what is typically done by other innovation spaces, Communitech purposely anchored its physical design, activities, and mission in values deeply rooted in the culture of the region, the values of collaboration, mutual support, and innovativeness.

This value-driven approach, combined with the strategic use of an associative model, the leadership and inspiring rhetoric of its CEO and key leaders, the anchoring of the organisation and its community in a physical place, as well as its extensive programming all contributed to its growth in notoriety, relevance, influence, and impact. It allowed it to attract significant support from public authorities and large corporations, and to better achieve its mission of orchestrating the development of the regional entrepreneurial ecosystem.

In part from the role played by Communitech, the development of Kitchener-Waterloo has been nothing short of phenomenal with hundreds of new technology start-ups formed, raising more than C$1 billion in investment, and creating tens of thousands of jobs. The importance of Communitech transcends the Kitchener-Waterloo borders and has become a symbol of Canadian entrepreneurship. Today, the region is considered by many as "the Silicon Valley of the North", one of the country's (and the world's) best innovation ecosystems, and an example to follow.

But in the face of such positive momentum and success, several questions come to mind. What could happen next? What does the future hold for Communitech and its entrepreneurial community? And now in a context of unprecedented public health concerns related to physical proximity, how should their physical place strategy evolve?

6.4.1 Facing important transformations

Communitech is undergoing important transformations. Its long-time focus on start-ups recently migrated towards scale-ups. As its community grew, it attracted other organisations that were created or got involved to complement (or compete with) their efforts and community building in more specialised areas and focus. Iain Klugman, Communitech's CEO, recently stepped down (in April 2021 after 17 years of service), to be replaced by an experienced executive in the entrepreneurial world, Chris Albinson (Communitech, 2021). Moreover, the Covid-19 pandemic, which pushed all start-ups and enterprise staff towards teleworking, has raised new and yet unanswered questions about the proper ways to connect communities and the role of physical places in general and the Communitech Hub in particular, in the future. While there is no doubt these changes are tackled with great planning, intelligence, and care, and that the organisation's leaders have remarkable track records and skills, these transformations are significant and could have important implications for the momentum of Communitech and of the regional entrepreneurial ecosystem.

Will the existence of a vibrant open lab, its formal structures, associative model, and physical hub, give Communitech the ability to adapt and maintain relevance on an ongoing basis? Will its distinctive value-driven approach continue to make a difference? Will this translate into more staying power? To what extent will the strategy of anchoring the organisation and its open lab in a physical place be relevant to the future phases of ecosystem development? Several other important questions must now be taken into consideration to ensure the resilience of this exemplary open lab organisation, so it can continue to drive the momentum of a lighthouse entrepreneurial ecosystem for years to come.

References

Architectural Record. 2011. The tannery. *Architectural Record*, February 15. Accessed 11 August 2021. https://www.architecturalrecord.com/articles/7385-the-tannery.

Bathelt, H., Kogler, D. and Munro, A. 2011. Social foundations of regional innovation and the role of university spin-offs: The case of Canada's technology triangle. *Industry and Innovation, 18*(5), 461486.

Bergek, A. and Norrman, C. 2008. Incubator best practice: A framework. *Technovation, 28*(1–2), 20–28. https://doi.org/10.1016/j.technovation.2007.07.008.

Bramwell, A. 2008. L'université, clé de la compétitivité du cluster TIC de Waterloo. *Le journal de l'école de Paris du management, 70*(2), 31.

Bruneel, J., Ratinho, T., Clarysse, B. and Groen, A. 2012. The evolution of business incubators: Comparing demand and supply of business incubation services across different incubator generations. *Technovation, 32*(2), 110–121.

CBC News. 2019. Waterloo region tech leaders predict change on new year horizon. *CBC News*, January 2. Accessed 30 April 2021. https://www.cbc.ca/news/canada/kitchener-waterloo/waterlooregion-tech-sector-predictions-1.4963521.

CBRE. 2020. Toronto remains Canada's top market for tech talent as smaller cities make gains during pandemic. *CBRE*, November 17. Accessed 30 April 2021. https://www.cbre.ca/en/about/mediacenter/toronto-remains-canadas-top-market-for-tech-talent-as-smaller-cities-make-gains-duringpandemic.

Cohen, S. 2013. What do accelerators do? Insights from incubators and angels. *Innovations: Technology, Governance, Globalization, 8*(3), 19–25.

Communitech. 2016, 2017, 2020. Reports, Communitech. Accessed 30 April 2021. https://www.communitech.ca/about-us/media-reports/.

Communitech. 2021a. Our story, est. 1997. *Communitech*. Accessed 30 April 2021. https://www.communitech.ca/about-us/our-story.html.

Communitech. 2021b. Digital main street. *Communitech*. Accessed 30 April 2021. https://www.communitech.ca/about-us/our-operation/digital-main-street.html.

Communitech News. 2017. Communitech @20: Two decades in, this is where we're headed. *Communitech News*, November 2, Accessed 30 April 2021 https://news.communitech.ca/communitech-20-two-decades-in-this-is-where-were-headed/.

Dingman, S. 2015. Startup city: The high-tech fever reshaping Kitchener-Waterloo. *The Globe and Mail*, July 17. Accessed 30 April 2021. https://www.theglobeandmail.com/technology/kitchenerwaterloo-startup/article25558263/.

Drori, I. and Wright, M. 2018. Accelerators: Characteristics, trends and the new entrepreneurial ecosystem. In M. Wright and I. Drori (Ed.), *Accelerators: Successful venture creation and growth* (pp. 1–20). Edward Elgar Publishing Ltd, Cheltenham, UK; Northampton, MA.

Gillmor, D. 2012. The invention of Waterloo. *The Walrus*, January 12. Accessed 23 March 2021. https://thewalrus.ca/the-invention-of-waterloo/.

Goswami, K., Mitchell, J. R. and Bhagavatula, S. 2018. Accelerator expertise: Understanding the intermediary role of accelerators in the development of the Bangalore entrepreneurial ecosystem. *Strategic Entrepreneurship Journal*, *12*(1), 117–150.

Grandadam, D., Cohendet, P. and Simon, L. 2013. Places, spaces and the dynamics of creativity: The video game industry in Montreal. *Regional Studies*, *47*(10), 1701–1714.

Grimaldi, R. and Grandi, A. 2005. Business incubators and new venture creation: An assessment of incubating models. *Technovation*, *25*(2), 111–121.

Imai, R. and Ban, M. 2016. Disrupting workspace: Designing an office that inspires collaboration and innovation. *Ethnographic Praxis in Industry Conference Proceedings*, *2016*(1), 444–464.

Klugman, I. 2015. Lean, agile and engaged: The Communitech corporate innovation model. *Tech News*, October 08. Accessed 30 April 2021 https://www.communitech.ca/technews/lean-agile-and-engaged-the-communitech-corporate-innovation-model.html.

Mérindol, V. and Versailles, D. W. 2016. Les laboratoires d'innovation ouverte comme dispositif entrepreneurial. *Entreprendre & Innover*, *31*(4), 52.

Moultrie, J., Nilsson, M., Dissel, M., Haner, U.-E., Janssen, S. and Van der Lugt, R. 2007. Innovation spaces: Towards a framework for understanding the role of the physical environment in innovation. *Creativity and Innovation Management*, *16*(1), 53–65.

Oksanen, K. and Ståhle, P. 2013. Physical environment as a source for innovation: Investigating the attributes of innovative space. *Journal of Knowledge Management*, *17*(6), 815–827.

Pauwels, C., Clarysse, B., Wright, M. and Van Hove, J. 2016. Understanding a new generation incubation model: The accelerator. *Technovation*, *50–51*, 13–24.

Peters, L., Rice, M. and Sundararajan, M. 2004. The role of incubators in the entrepreneurial process. *The Journal of Technology Transfer*, *29*(1), 83–91.

Roundy, P. T. 2021. Leadership in startup communities: How incubator leaders develop a regional entrepreneurial ecosystem. *Journal of Management Development*, *40*(3), 190–208.

Schmidt, S. and Brinks, V. 2017. Open creative labs: Spatial settings at the intersection of communities and organizations. *Creativity and Innovation Management*, *26*(3): 291–299.

Spigel, B. and Bathelt, H. 2019. Questioning cultural narratives of economic development: An investigation of Kitchener-Waterloo. *The Canadian Geographer/Le Géographe canadien*, *63*(2), 267–283.

Spigel, B. and Stam, F. C. 2016. Entrepreneurial ecosystems. *USE Discussion Paper Series 16.13*.

Stam, E. 2015. Entrepreneurial ecosystems and regional policy: A sympathetic critique. *European Planning Studies*, *23*(9), 1759–1769.

Stam, E. and Welter, F. 2020. *Geographical contexts of entrepreneurship: Spaces, places and entrepreneurial agency*. Working Paper, No. 04/20 (Institut für Mittelstandsforschung (IfM) Bonn, Bonn).

Startup, G. 2020. The global startup ecosystem report. GSER 2020, Startup Genome. https://startupgenome.com/all-reports.

Steelcase. 2015. Innovation center ideabook. Accessed 30 April 2021. https://www.steelcase.com/content/uploads/2018/08/innovationcenterideabook.pdf.

University of Waterloo. 2021. Policy 73 – Intellectual property rights. Accessed 30 April 2021. https://uwaterloo.ca/secretariat/policies-procedures-guidelines/policies/policy-73-intellectualproperty-rights.

University of Waterloo. 2016. "UWaterloo's Velocity Garage now largest free startup incubator in the world. " Accessed 30 April 2021 https://uwaterloo.ca/news/news/uwaterloos-velocity-garage-nowlargest-free-startup.

Welter, F. 2011. Contextualizing entrepreneurship-conceptual challenges and ways forward. *Entrepreneurship: Theory and Practice*, *35*(1), 165–184.

7 Building communities in rural coworking spaces

Ignasi Capdevila

In the context of a rising sharing economy and the growing number of freelancers and knowledge workers, the last two decades have witnessed the worldwide spread of the phenomenon of coworking (Gandini, 2015; Merkel, 2019; Spinuzzi, 2012). The main difference between coworking spaces and shared offices is that coworking is characterised by collaborative dynamics and the central focus of knowledge sharing. Being able to infuse these values and to create a sense of community (Blagoev, Costas, and Kärreman, 2019; Garrett, Spreitzer, and Bacevice, 2017; Rus and Orel, 2015) is thus crucially to succeed in creating a collaborative working atmosphere (Capdevila, 2015; Liimatainen, 2015).

So far, the vast majority of this academic research and policy interest in the phenomenon of coworking has focused on forms of urban activity, as a way to nurture collaborative local networks around focal interests. Over the past decade, an increasing number of academic and policy interventions beyond metropolitan centres have nonetheless begun to consider coworking in rural and peripheral areas. Some initiatives have been developed, suggesting that coworking in rural areas could not only benefit coworkers, but also contribute to support local policies aiming to develop the socioeconomic context. In this sense, rural coworking appears often as a public service, contrasting with some urban coworking that exploits real estate business models and the commoditisation of communities.

Considering the lack of research centred on rural coworking, and the rising interest from policymakers to understand its potential, this chapter focuses on understanding one of the main aspects of coworking, and probably one of the most complex to develop: a lively and collaborative community. With this aim, the chapter presents how the different spaces associated in the Cowocat Rural (the network of the rural coworking space in Catalonia) struggle to identify potential coworkers and to support the development of communities, at the level of the space and at the municipal level, as well as at the county and regional levels.

7.1 From urban to rural coworking

Coworking spaces are shared office environments where a heterogeneous group of independent workers pays (generally a monthly fee) to use as their

DOI: 10.4324/9781003125587-10

place of work, to engage in social interaction, share assets and services, and collaborate on projects of mutual interest. Coworkers are generally freelancers and independent workers even though in some cases they are teleworkers working for an organisation. Coworking spaces are generally operated by private actors (ranging from small firms to large real estate corporations), although sometimes they are run by associations or non-profit organisations. In its origins, coworking has been an urban phenomenon and is commonly found in cities, where the density of work interactions and economic actors, facilitate the development of digital entrepreneurial and freelance work modes, characterised by contacts with multiple actors and high degrees of uncertainty.

Coworking services are usually presented as "joining a community" (Gandini 2015; Blagoev, Costas, and Kärreman, 2019; Rus and Orel 2015). The principles of coworking imply high social interactions that are generally enabled by the action of space managers that act as curators (Brown 2017) to facilitate personal introductions, organise events and training activities, as well as developing and maintaining active contact networks (Parrino,2015). In this sense, coworking spaces represent community-based organisations based on mutual help and collaboration, a key aspect in the emergence of innovations. These new organisations have also been referred to as open labs.

Coworking spaces allow members to share a physical space and be in frequent physical proximity, thus enhancing knowledge sharing and learning. Other forms of proximity, like cognitive and social proximity (Micek 2020; Mariotti and Akhavan, 2020) also contribute to the development of a coworking "spirit" (Avdikos and Iliopoulou, 2019) or a sense of community (Garrett, Spreitzer, and Bacevice, 2017) that is fundamental to the development of trust and social bonding to enhance mutual help and collaboration.

In the evolution of coworking, different stages can be identified, corresponding to different types of spaces in terms of ownership, intentionality, and strategic focus. Two main categories can be distinguished.

On the one hand, there are private for-profit spaces, that intensively focus on developing business opportunities for their members. These spaces have been referred as entrepreneurial-led coworking spaces (Avdikos and Iliopoulou, 2019) or neo-corporate spaces (Gandini and Cossu, 2019). They generally adopt a top-down governance mode, where the relationships are based on contractual agreements of space renting, and an offer of services for a monthly fee. Even if coworkers have a strong influence in the organised activities, the decision-taking processes are mainly top-down. This coworking practice is rather linked to the commodification of work and the flexible labour dynamics characteristic of the gig economy (de Peuter, Cohen, and Saraco, 2017; Gandini, 2015).

On the other hand, there has been a more recent trend on spaces that are community-based types of organisations and rather focus on social and interactional aspects, refereed as community-led spaces (Avdikos and Iliopoulou, 2019). These spaces are characterised by being based on a bottom-up initiative originated by a self-organised community of professionals, generally related to a certain domain or industry, that jointly rent and share a workplace, share facilities and assets but, most importantly, share knowledge and engage in collaborative practices fuelled by mutual aid and trust. In that perspective, coworking represents a response to precarity (de Peuter, Cohen, and Saraco, 2017)

Even if extreme cases of both types of spaces can be distinguished, there has been a progressive convergence of the two models, and nowadays many spaces present characteristics of both. The emerging coworking industry based on real estate companies such as Regus or WeWork have elaborated communication strategies based on terms like community and trying to capitalise on their customers' collaborative dynamics. At the same time, many community-based spaces have adopted more structured decision-making processes and have taken a more business-oriented approach. However, both trends are still identifiable through differences in their governance, design and characteristics of their physical space, or the organised activities.

In terms of the temporal evolution of coworking, three phases can be distinguished (Gandini and Cossu, 2019; Avdikos and Iliopoulou, 2019; Gandini, 2015). The first wave was mainly constituted by an urban practice that aimed to counter the increased precarity and uncertainty encountered by freelancers in an ever-increasing commodification of the job market (Merkel 2019; de Peuter, Cohen, and Saraco, 2017). In the early 2010s, coworking spaces quickly expanded in major cities. Most of them were private initiatives of professionals who needed to share costs and break the social isolation of working from home. Coworking practices started to be developed around community-building activities among small groups of peer workers. As the coworking principles gained adopters, the movement increasingly expanded, with literally hundreds of spaces being opened every year in main urban areas across Europe and other main hubs around the world. The real estate market identified and opportunity of monetising empty offices (in many cases as a result of the global financial crisis of 2007–2008, like in the case of Barcelona), and started to adapt their underused offer to respond to more flexible and independent work modes. Large firms developed a model of "coworking as a service" for freelancers and firms desiring to reduce their renting costs, externalise services, and increase work flexibility. This "corporatisation" of coworking represented the second wave of the expansion of the coworking concept, that contributed to making coworking a global and visible phenomenon. This new image resonates with the open-space offices of large high-tech companies like Google or Facebook, and with the entrepreneurial atmosphere of Silicon Valley. The neoliberal and corporate turn was criticised, arguing that it was somehow denaturalising the initial values of coworking (de Peuter, Cohen, and Saraco, 2017; Gandini and Cossu, 2019). This new corporate focus based on a "paying to work" paradox, also reinforced discourses defending the communitarian bottom-up essence of coworking, that many spaces claim (Merkel 2019; Avdikos and Merkel 2020). The response to commercial coworking has been identified as a third wave of coworking, and it has been identified by the emergence of a type of space that blends the entrepreneurial aspects with a social orientation. Gandini and Cossu (2019) have referred to these spaces as "resilient spaces", seen as a countermovement to the mainstream neo-corporate coworking model. This third wave has also implied a decentralisation of the phenomenon, that until then, was almost exclusively located in urban areas. Starting from the early 2010s, rural or peripheral areas saw the launch of collaborative spaces, initially following the urban model (Capdevila, 2021). In contrast to spaces in metropolitan areas, spaces in rural areas are often developed as ways to enhance the local economy as well as to support entrepreneurship and the new professional perspectives for local youth, that is often tempted to migrate to nearby cities. Opening spaces in deprived regions is often considered as a policymaking tool to structure local professional networks and to provide services to independent workers (Fuzi 2015; Flipo, 2020). Consequently, in many cases, the spaces are public, even if they coexist with other private initiatives (Fasshauer and Zadra-Veil, 2016). Rural coworking spaces are very much embedded in their local environment, as they become "social hubs that deliver a number of wider social services to the local communities, while they are more attached to them" (Avdikos and Merkel, 2020: 354).

Rural coworking has recently raised the interest of municipal councils and policymakers, and its development has received funding from the European regional development agencies. However, it is still in an incipient phase of development and its impacts on local economies and on the working conditions of rural entrepreneurs and freelancers are still unclear. Nevertheless, its potential, as a way to help in talent retention (and even talent attraction, mainly after the forecasted urban exodus caused by the Covid pandemic crisis), as well as springboards to enhance local collaborative networks has interested regional policymakers to support initiatives and closely track their evolution.

7.2 Building communities in coworking spaces

The different approaches, discourses, and practices about coworking have a common trait: the revindication of the importance of a community. Community-building is one of the main characteristics of coworking, and it is the main difference between coworking and office sharing. Community-building, even if underlined as a constitutive element of the practice of coworking in both rural and urban areas, might be referring to a different phenomenon in each case. In the case of research on urban coworking, the term community has been considered as an umbrella concept with blurred connotations. Some authors have tried to disentangle the complexity of the term "community" by conceptualising it in a social network approach (Liimatainen, 2015), using Bourdieu's concept of habitus to the symbolism of community (Butcher, 2018) or Adler and Heckscher's (2006) typology of communities based on the concepts of *gemeinshaft* and *gesellschaft* (Spinuzzi et al., 2019). Rus and Orel (2015) further developed the concept and proposed a decomposition of the concept into four aspects: functional, structural, cultural, and territorial. Communities in coworking spaces combine a social orientation with a business focus, assimilating their dynamics to community-based organisations (Mérindol, Aubouin, and Capdevila. 2021). While studying urban spaces, some approaches have been more critical, and have claimed the artificiality of the concept and its positive connotations that not always reflect a collaborative reality, but rather a corporate discourse (de Peuter, Cohen, and Saraco, 2017). In this line, Gandini (2016) argues that communitarian relations reproduce fictitious institutionalism that permits the marketing of subjects. In a similar vein, Jakonen et al. (2017) claim that the concept of community is commodified, losing its essence.

In the scarce literature on rural coworking spaces, community is generally understood as a wider term, including not only the core members of the space but also the larger community of its geographical context. Different layers of community can be distinguished, ranging from the internal community to a larger understanding, including external actors with more or less strong ties to the core community. Research on urban coworking has generally focused on the internal community dynamics, stressing the benefits of working among peers, like facilitating knowledge sharing, developing trust, peer-learning, and collaborating (Spinuzzi, 2012). The coworking literature has defined urban spaces as "collaborative work environments" (Fuzi, 2015) often referring exclusively to the core community. Parrino (2015: 265) affirms that "the concept of community refers to the possible relational implications of the co-location of workers within the same space and emphasises the role of coworking as a work context able to provide sociality to coworkers". In great part, intensive collaborative dynamics appear as a consequence of shared interests and professional specialisations, characteristics of urban spaces (Blagoev, Costas, and Kärreman, 2019). In opposition, in rural areas, where the low density makes the concentration of professionals with a common specialisation difficult, collaboration rather than being topic-oriented is more based on searching for complementarities (Avdikos and Merkel, 2020).

Community-building is developed in two different but complementary ways. On the one hand, informal social interactions among members naturally create bonding, based on trust, that leads to a sense of community developed through frequent interaction (Garrett, Spreitzer, and Bacevice 2017). Leclercq-Vandelannoitte and Isaac (2016: 6) argue that "coworkers tend to co-create a sense of community" that they understand as being more genuine than a top-down imposed concept of community. On the other hand, community-building can also be fostered by intentionally creating favourable contexts of interaction and collaboration, such as the organisation of events and social activities, that gather the members together, thus creating opportunities for socialisation and knowledge sharing. These activities, even if they can be proposed by the members, are generally organised by the space managers. Community-building then also appears as a purposeful set of actions that take place in coworking spaces (Cabral and van Winden, 2016). The literature in urban coworking has reflected the interest of space managers in the internal synergies that events create, and it has studied, to a smaller extent, the impacts on the local environment. Advocates and promoters of rural coworking have nevertheless been interested in the external effects of coworking for the territory surrounding the space, as a way of improving the social integration at the local level, contributing to the local revitalisation, in economic and spatial terms, and the strengthening of social ties and the sense of a local shared identity (Akhavan et al., 2019).

So far, the study of the communities in coworking spaces has been mainly based on urban areas and, as the empirical results show, rural and urban coworking practices are different in various aspects that are worth analysing. Also, the different characteristics and aims of urban and rural coworking might mean that the impact of communities on their local environment might differ, so research results from urban coworking might not be able to explain realities in rural environments (Akhavan et al., 2019; Capdevila, 2015). Therefore, the goal of this research is to focus on the communities in coworking spaces in rural areas, focusing both on the internal and on the external spatial dynamics.

7.3 Methodology

This chapter draws upon a five-year study of coworking in Barcelona and Catalonia. Data were collected in three phases. The first phase, from 2013 to 2015, consisted of a research project in the coworking spaces in Barcelona, including 28 interviews with managers and coworkers. Data triangulation was done with other nine interviews of academics, policymakers, and specialists highly knowledgeable of the Barcelona innovation dynamics.

In a second phase, from 2015 to 2017, a new round of semi-structured interviews was conducted with 27 managers and coworkers of eight Catalan rural spaces, as well as representatives of the Catalan government, regional coordinators, and managers of Cowocat Rural.

In a third phase, seven follow-up interviews were done with the coordinators of the Cowocat Rural, representatives and some of the spaces' managers, with the goal of having a longitudinal understanding of the evolution of the network of rural spaces.

Finally, archival data (like materials for presentations and for managers' training, established criteria for selection of coordinators, guidelines to managers, etc.) and online content of the different spaces were used to understand the different stages in the development of coworking in rural areas.

To analyse the data from a community perspective a number of milestones were identified, including the progressive opening of rural spaces, along with the evolution in the discourses around coworking from policymakers and managers of Cowocat Rural. That first analysis sought to identify the progressive implementation of rural coworking in Catalonia. Second, the analysis focused on gaining a deeper understanding of the role of the different actors (municipal, regional, and government policymakers, space managers, and coordinators of the Cowocat Rural network) in the construction, and management of the communities emerging in or around coworking in rural areas. Third, the different levels of community were characterised for each space and then analysed through a comparative case study.

7.4 Research context: the case of Cowocat Rural

Over the last decade, Barcelona has become one of the most important European hubs for coworking (Coll-Martínez and Méndez-Ortega, 2020; Institut Cerdà/AMB, 2019). There were over 100 spaces in Barcelona using the term coworking to define themselves (Capdevila, 2015), if the number is highly variable, and likely to be strongly affected by the effects of the Covid-19 crisis. From the 2010s, public bodies (notably the Catalan government) started to be interested in coworking, considering it a way to boost entrepreneurship and to help unemployed people to join professional work settings. The Catalan government promoted the creation of Cowocat, to capitalise best practices and to create synergies among the spaces. This network of coworking spaces facilitated the diffusion of the practice of coworking in Catalonia, raising awareness and interest among town councils from rural areas attracted by the idea that coworking could contribute to retain talent and develop the local economy.

In 2011, in Riba-Roja d'Ebre (a small village 200 km south of Barcelona), a group of young councillors had started to invest in a local fibre optic network while considering the possibility of rehabilitating the former public library into a coworking space. They got in contact with technicians of the local consortium of socioeconomic initiatives "Consorci Intercomarcal d'Iniciatives Socioeconòmiques" (CIS) to assist them in determining the technical aspects of such a space. By that time, the manager and one technician of CIS were already interested in the concept of coworking gaining popularity in Barcelona. The CIS team studied the urban coworking phenomenon to adapt it to rural areas. In 2012, the coworking space (called Zona Líquida) was launched through the development of the "Consell Comarcal" (County Council – CC). The goal of the pilot project was to revitalise the territory and to strengthen the relationship among local entrepreneurs.

The project, which started as a public space open to local entrepreneurs, was often empty due to a lack of targeted communication and a certain difficulty in identifying potential users. In the beginning, the experience did not achieve the expected results. To try and engage local entrepreneurs, the promoters selected five young rural entrepreneurs and freelancers (according to their professional profile) and sent them on a trip to Barcelona to visit five different coworking spaces. They came back being convinced of the coworking "philosophy" and the importance of community management activities that had to be organised by dedicated persons. Following their recommendation, Zona Líquida started to open on a daily basis, and hired a community manager, responsible for the search for new members, and their integration by facilitating networking among members. After being almost empty for the first two years, professionals from the village and surroundings progressively started to use the space.

Then, in 2014, the CIS and CC thought that the initiative could be replicated in other Catalan rural areas and started a project open to other partners with the aim of retaining talent, attracting population, and helping local entrepreneurs. That year, the project Cowocat Rural was launched, with the financial contribution of EU rural development funds. The project started as a network of rural coworking spaces affiliated to a larger network (Cowocat, which mainly included Barcelona-based coworking spaces). Cowocat Rural is a non-for-profit initiative with support from public institutions, notably the Catalan government, and currently consists of 34 rural spaces, 16 of them publicly funded and 18 private.

The promoters of Cowocat Rural encouraged coworking by offering support to village councils looking to open a space. One of their main challenges has been to explain that the success of coworking depends on the pre-existence of a community. The characteristics of rural areas, with low population densities and high transport times, make the frequent co-location of members in the same space more difficult than in the urban environment. That is why the spread of coworking is not only done by opening new physical spaces, but also, and more importantly, by previously analysing the territory, its economic and industrial trajectory, and its current needs and by developing the local community. In particular, the identification of existing professional communities was considered as an important precondition before taking the decision to open a space.

A commonly identified mistake in rural coworking has been to rush into the launch of a space without considering the current needs or the potential users. Given the lack of knowledge about the values and principles of coworking, promoters engaged in "evangelisation", both at the level of citizenship and at the institutional level, by providing examples of real successful cases.

The project has since been extended with several side projects. For instance, an initiative called "Rural & Go" proposed temporary gatherings of one to two weeks to experience co-living in rural environments to urban coworkers. The "Rural Pass" programme allows coworkers to work punctually in other spaces in the network. The promoters also launched a "coworking lab" in a university, where researchers analyse the collaboration dynamics in real conditions. Also, there was an initiative to develop an online platform to facilitate the interaction and collaboration among rural coworkers located in different spaces in Catalonia.

7.5 Results

This section presents the results, firstly, by describing the different levels of the communities, secondly, by relating these different levels to the type (face-to-face or virtual) and intensity (frequent or occasional) of the interaction among members, and, thirdly, by analysing the main challenges that promoters encounter in the diffusion of coworking in rural areas.

In the Cowocat Rural case, several layers of communities can be distinguished (at the level of space, at the level of municipality/county, and at the regional level), depending on whether the links between members are strong and interaction is frequent and co-localised, or whether the relationship is more sporadic and distant.

7.5.1 Community at the level of the physical space

The first level is the community in the coworking space. It is composed of the members who come to the space most frequently (almost daily) and for whom the space represents its usual working place. In the coworking spaces of Barcelona, this community can gather dozens of people. In the case of rural coworking, these communities usually have an average of six coworkers. The social bonds that are created are strong because the interaction is constant, which often creates a high degree of mutual trust. In these cases, coworkers often collaborate and help each other, although sometimes they are exchanges of favours or billing of services below market price. These small communities (ranging from three to ten members) also represent a moral and personal support network, which goes beyond the purely professional.

As detailed below, the existence of a community (or at least a group of people potentially interested in sharing a workspace) prior to the opening of a coworking space ensures that, once operational, the space will have certain initial interactions among new members and an optimal context for collaboration. In some municipalities, where the opening of a space has been prioritised instead of a previous identification of interested people, the development of a community represents a greater difficulty for the success of the project.

The example of Riba-Roja illustrates the composition of the core of the community and of a wider local community in the daily use of the space:

> We are the three who come most often. Then there is one person that comes a few times a week, like the Englishman. There are three more people who come two or three times a week. Or the meeting room, where courses are held. It is also used by associations.
>
> (Coworker – Zona Líquida)

7.5.2 Community at the municipal and county level

The community of frequent members is complemented by a second type of community, which integrates people who come to the space more sporadically. Some spaces have shared tables, where coworkers that come from time to time can use.

> In my territory, today, I find it very difficult to have a space with 15 tables and all full of the same people every day. I rather see a space with 15 tables

> and three or four of them every day with the same people, but with a lot of rotation.
>
> (Space manager – Noguera – Segrià Nord)

It is worth mentioning that in spaces where no monthly fee is charged (often the case of public spaces owned and run by the municipality), sporadic use is facilitated, increasing the interrelationship and contact between professionals living and working in the area.

To keep this community alive and informed, some spaces choose to create an online directory, where all registrants are informed about news and scheduled events.

> We've realised that these people coming and going, need to punctually locate themselves in one place. On our website, we have opened a section that is for the community. All the non-coworking people [*who don't come on a regular basis*] we list them there. To give talks, etc. We have them registered. They feel that they belong to the community. At one point, we had a chat, and we know they're there. As a coworking community we are not enough people to give talks just for us.
>
> (Space manager at Coworking Alfarràs, Alfarrràs)

> In the space, we are now ten people. Besides, there is an English teacher who comes to teach his classes here, the yoga teacher, and there is a community of seven or eight people who are not here but are in the directory, they are informed of everything we do. They are part of the community without having a fixed space.
>
> (Space manager at Espai La Magrana, Valls)

Compared to urban coworking, rural coworking suffers from a critical mass problem. In a small community, it will be more difficult for different professional profiles to find a certain specific affinity. Instead, the probabilities of synergies are multiplied by considering a wider community. As in other community aspects, the role of the facilitator is paramount. It is therefore important that the manager knows first-hand the profiles and needs of the coworkers.

> But the important thing is that even if [the one-time members] don't come, they are within the community. They may not come every day, but they will come if there is a good manager who offers them meeting points: a specific training, a meeting that interests everyone, etc.
>
> (Network manager, Ribera d'Ebre – Terra Alta)

These larger communities also include people who have come to an event in the past or former members who, in many cases, remain linked to the spaces.

In the case of the private spaces Coworking Alfarràs or Espai la Magrana, in Valls, communication with the community is done through an online platform. Visibility is also enhanced by the brochures and business cards of the different coworkers at the entrance of the venue. In the case of La Magrana, they have organised a network at the Camp de Tarragona regional level to facilitate collaboration between spaces of the region.

In the case of the Salines-Bassegoda consortium, the idea is to create a community around a few spaces:

> Within Salines-Bassegoda, which are the 17 municipalities, we believe that there may be between four and five [coworking spaces]. People will not travel 30 or 40 minutes, which is what it takes between one village and another. We believe in creating a community with different spots. In each spot there can be five or six fixed people using the space, and then create a community of 30 or 35 who can work together.
>
> (Space manager, Salines-Bassegoda Consortium Technician)

7.5.3 Community at the regional level

Apart from the online networks of the spaces, the Cowocat Rural project has launched the development of an online platform where different coworkers can register as long as they are members of a space. Coworkers can, for example, create projects and disseminate them publicly to solicit collaborations. In this respect, a first project has resonated in another space:

> I introduced a sustainable gardens project, and just yesterday I received an email from Navata coworkers, who were landscaping architects who were contacting me to see if we could collaborate. This is very beneficial.
>
> (Coworker and facilitator, Riba-Roja d'Ebre)

The website of the Cowocat Rural network, apart from the synergies among affiliated coworkers, allows the relationship between the facilitators to share their experiences, successes, and failures. As a manager states: "You have to be in touch with the network, to learn from the mistakes of others. At first, you don't have much idea, but the others know about activities and so on. Others have already tried it. Don't make the mistake again".

7.5.4 Face-to-face or virtual interaction? The issue of travel time

The phenomenon of coworking is generally associated with physical spaces and face-to-face interaction. But beyond physical space, the philosophy of coworking is, above all, a way of working based on collaboration and the search for synergies among community members. Face-to-face meetings, whether frequent or one-off, strengthen the bonds between people, but the coworking philosophy also has a place in virtual relationships of collaboration and mutual help. Virtual relationships in communities are even more important in the rural environment, where the distance and transportation times might dissuade many people from going to a coworking space often. A fact that is often detected by managers interviewing potential coworkers is that people in rural areas are somewhat reluctant to travel more than 10 or 15 kilometres to a coworking space.

> In Barcelona people are able to move an hour by underground train, but offering this service to people from Móra, who would love to cowork will never come to you because: "oh, half an hour to go and half an hour to return, I don't care". Many people from [nearby villages] have told me. People from Gandesa who initially wanted to come once a week to make contacts and have not ended up coming because they are wasting their time driving.
>
> (Coworker and facilitator, Riba-Roja d'Ebre)

The climatic factor also plays an important role. In winter, in certain territories, transportation is hampered by fog, ice, or snow. These obstacles to mobility make spaces devise various strategies. One of them is the atomisation of spaces and the creation of local networks composed of small spaces. In the case of Salines-Bassegoda, for example, the territory consists of 17 municipalities. Travel times make it difficult to create a set of four or five spaces in order to satisfy a maximum of people and optimise transport times. In this case, the communities of each space will be small, but the idea is to organise joint activities to revitalise a larger community of about 30 people.

Another strategy is based on the articulation of different types of interactions, some face-to-face and others virtual. Table 7.1 summarises it.

As discussed above, people who use the space as a regular workplace constitute the core community. Core members, especially if the space manager is part of it, are usually the ones that interact virtually through social media and who upload information about activities, courses, or the agenda of events in the website. Then, there are the people who regularly use the space and are just looking to build bonds with the other regular members. These first two groups form what we have called the community at the level of the space. A third group consists of people who regularly follow the information about the

Table 7.1 Face-to-face and virtual interactions in rural coworking spaces

	Frequent face-to-face relationship	*Occasional face-to-face relationship*
Frequent virtual relationship	Managers and involved members of the core community of the space	Extended community: involved mainly in contact via social media and occasional participation in events in the space
Occasional virtual relationship	Core community of the space	Sporadic relationship with the community

space on the internet but who come on a one-off basis, either to participate in a specific activity or to work occasionally. This group is what we have called community at the municipal or county level, interested in the community but with difficulties in participating regularly. Finally, there are people who sporadically follow space activities, either through social media or a visiting physical space from time to time. These people can be both from outside the territory or present in the territory but have a relatively distant interest in participating in the space activities.

The risk to the core community is that it stays closed on itself, limiting exchanges with outsiders. That is why it is important that community managers continuously make efforts to promote the space and attract new members. The richness of the collaborations will strongly depend on the diversity and number of different profiles that will be contacted. Not only relationships based on strong bonds are constructive. In fact, the communities can capture a lot of value derived from relationships with external individuals with whom they have weak ties. That is why it is important to keep in touch with the four types of groups represented in Table 7.1, without underestimating the peripheral members in the community, such as, for example, coworkers from other spaces in the Cowocat network.

7.5.5 Identification and support of existing communities

One of the most important aspects of the previous analysis is the identification of entrepreneurs and self-employed professionals present in the territory, as well as understanding their needs. Also, to study their distribution in the territory (to understand if they are distributed or concentrated in specific areas) and their organisation (for example, identifying local groups or associations).

Understanding the reality of local entrepreneurship and the self-employed population is important to ensure that coworking will meet their needs. The network managers identified as important the fact that, before considering the opening of a physical space, efforts are made to identify, unite, and strengthen the professional community on which the project is focused. This implies that the activities do not have to be isolated and punctual in time, but it is an evolutionary process within the development of a local community. This process must also be iterative and adapted to the progressive evolution of the initiative:

> There has to be a whole lot of work and reflection. A research perspective has a lot to do with it. We don't go impulsively but we look, we take a few steps, and see how we should go on. Many times, if this reflection is not done, the effect is reversed. We can make an investment and if it does not produce anything, we have the opposite effect, that the citizens turn against it. "Much money has been spent on this …". The effect is the opposite; it is the worst you can have. It will kill any initiative for years.

> People remember the failure. There must be a certain rigor when it comes to seeing how to do this type of action in a territory.
>
> (Head of the Digital Inclusion and Training Service, Catalan government)

In the most successful cases, promoters have considered as a priority the identification of the target group before the opening of a physical space. At the same time, when implementing the strategy, the fact of not having a space can make it difficult to disseminate the project. So, very often the dilemma that arises is whether it is necessary to have a community of interested people as a precondition to open a physical space, or if it is necessary to have a space to attract professionals. Table 7.2 summarises the advantages and disadvantages of identifying a community, before or after opening a space.

In general, the best practices identified have been about identifying a community before opening the space. According to a Cowocat Rural manager: "you have to look for [the community] and when you have created it, people get together, and they always end up looking for a space". You can energise a community of professionals without having a physical space, but you can't energise a space if you don't have an interested audience. In addition, having an empty space, apart from the expenses it represents, can be counterproductive for the negative image that it projects. Before considering opening a new space, identifying local profiles potentially interested in coworking represents one of the main tasks of the coordinators of the Cowocat Rural project:

> We meet them [potential members] to do an interview, we explained them what coworking is. At that time there was a space created. All this meant that, if they were interested, we could work in the space. Elsewhere, what we've done, in villages that don't have a space, the first thing we do is to start looking for profiles that can integrate a community and are on the territory. There might be people who are working from home. But they

Table 7.2 Identifying a community: before or after opening a space?

	Advantages	*Disadvantages*
1) Identify community → 2) Open a space	Understanding the real needs (feasibility study). Adaptation of the project to the needs of the community. The space is used from the start. Cost optimisation.	Lack of visibility and difficulty in disseminating the concept of coworking.
1) Open a space → 2) Identify community	Ease of communication and dissemination of the concept.	Risk of space being underused. Space not adapted to real needs. High costs. Risk of low profitability.

> do not know each other. But even if they know each other, they still end up in the municipal library [...]
>
> You can't go find an empty space and just open it. What we see is that you have to go and see if you have people in the territory to be able to make a space or a community. There are councillors or town councils who come to us and say: "I want to create a coworking space". Because there is also a grant. Without considering what the goal is, what it is for. [...] We think that a diagnosis must first be made to see if the need exists in the territory. We could have opened a space, but we saw it wasn't the best idea. Because if you open a space and put some tables and chairs, it doesn't fill up.
>
> (Network manager, Ribera d'Ebre – Terra Alta)

The strategy for identifying profiles potentially interested in rural coworking may differ from one initiative to another. In some cases, the initiative has been the result of a need expressed by a group (bottom-up dynamics). For example, in the case of the village of Tremp, the project arose from a series of surveys conducted by youth leaders and showed that they wanted a space to work in Tremp. In other cases, the initiative arises from a public or private body interested in testing the viability of such a project (top-down dynamics).

In general, the target professionals (whether entrepreneurs, freelancers, ICT professionals, SMEs, etc.) are not organised in local communities or associations, and project promoters need to actively search for profiles. A common first step is to contact the bodies responsible for economic promotion, youth, entrepreneurship, or the city council.

> We know that there are people who work online and have liberal professions. That is, they are freelancers and work from home. We get there from Economic Promotion and Youth [department]. Contacts in the territory. The youngest profiles reach Youth [department]. In Economic Promotion [department], there is a business incubator. They have a database of about 50 people and perhaps there are cases when they say, "this one has a more individual profile, and he could be in a coworking instead of being in a start-up accelerator". We also work a lot with the Ripollès [county] development agency to look for profiles. Or through word of mouth.
>
> (Technician of the Cowocat Rural project, Ripollès)

The rural environment also makes it possible to identify profiles through word of mouth and personal contacts, in a snowball strategy. For example, in the case of an individual initiative in Artesa de Segre, before looking for a place, the promoter looked for interested people through contacts and social networks:

> I made a list of people to call. Talking to people. For example, I know some women who are lawyers, and they live in a village. I know that they

> have an office in Tàrrega but they work more and more here in Artesa. I know some people that I already have targeted, and I think they might be interested. They are already three or four.
>
> (Coworker, designer, Artesa de Segre)

> I keep interviewing people from the two counties of Terra Alta and Ribera. Both Toni and I are looking for new users. Especially word of mouth. Contacts, contacts. Being a small town, you don't have any "yellow pages" [phone directory] or someone to tell you how many self-employed people there are. With the data protection law, if you go to a centre asking for a list of self-employed people in the municipality, they will tell you that they can't give it to you. So, it has been fieldwork, asking staff, people who are interested, through social media.
>
> (Coworker and facilitator, Riba-Roja d'Ebre)

The territorial perimeter that it is intended to cover must also be established, whether at the municipal or county level. The low population density in rural areas will tend to consider a larger area than the county. This also increases the chances of finding a critical mass of interested people and finding complementary and related profiles. In the case of Riba-Roja, for example, not all coworkers are from the village, or even from the nearby area. However, the territorial scope of the project will also depend on the existence of other coworking spaces within the same territory, on the critical distance (from which, people consider a space too far to move), and the need for new spaces.

Identifying a group of people initially interested is also no guarantee of success. Sometimes the initial intentionality is not reflected in the facts. For example, in the case of Tremp, although the request came from young people in the village, they did not make intensive use of it once the space was opened. The risk may also be that the implementation of the space or the end result will not meet the initial expectations.

7.5.6 Dissemination of the concept of coworking

The results show that one of the most important initial difficulties is to diffuse what coworking is and what it is for. The challenge is especially important at the beginning of the projects. In the event that the project begins with the opening of a space, interviewees underlined the importance of being based on the principles of coworking (sharing, collaborating, socialising, helping each other, etc.). The risk is that the first experiences condition the dynamics that are later established, and the project ends up losing the initial philosophy and becomes a social space or a shared office. The first few months are especially delicate, as the usual low number of coworkers can be interpreted as a negative sign and the project may lose support. The process and evangelisation of coworking and the recruitment of coworkers can be slow:

> In the first year, you begin to understand what is rural coworking in the area. People from neighbouring villages start coming, […] you begin to see a professional movement and people who show interest and ask to join.
>
> (Salines-Bassegoda Consortium Technician)

Institutional support is needed during the consolidation process of coworking, as ignorance about coworking and a slow progression can be a source of misunderstanding and criticism of the initiative for lack of tangible results.

> From a local point of view, [the goal is] to consolidate, to be seen as a real option for people to live and work in Riba-Roja, regardless of consolidated quantitative results. I quite refuse to quantify the success of projects on a numerical basis. Here we can't work like this because it goes against rural philosophy, but we do notice a kind of glass wall in relation to a part of the population that finds it difficult to understand what we do, they try but it is difficult, but they will better understand it when there will be more native profiles that have really been able to stay in the village.
>
> (Councillor for economic promotion of Riba-Roja d'Ebre)

One way to make coworking better known is to give more visibility to those who practice it, to open the space to neighbours so that they know the space and the people who work in it, to spread the concept, and at the same time, the professional services and activities that are offered.

> You have to think that, at first, we were like "weirdos", those who worked in the town hall. Nobody knew about us. After two years, people know we're coworkers, we have coffee together, and people know we work together, and we don't work for the city council.
>
> (Coworker and facilitator, Riba-Roja d'Ebre)

An additional problem is that the concept of coworking is generally related to urban coworking, and it is still associated with the activity of renting spaces and tables. From this perspective, critics of rural coworking argue that to have access to a workspace is not a problem in rural areas. As a manager explains: "In a rural area, there is always plenty of available tables and spaces. Everyone has their grandparents' house that is empty, or two warehouses: there is a lot of space". The challenge for rural coworking is often to explain the fundamental difference between urban and rural coworking, where the important aspect is to increase the interaction among professionals, rather than to reduce the costs arising from the sharing of spaces. If coworking is understood only as a cost reduction, then it does not provide any added value compared to other existing workspaces, such as business centres or incubators that already exist in the territory. As a space manager comments: "We have to go much deeper into coworking communities and revitalisation, rather than into renting spaces".

In fact, understanding this difference and managing the space accordingly, is a cornerstone of a space's success. For example, in the case of Flix and Riba-Roja, two villages separated by a few kilometres, the philosophy applied to the spaces (shared office in the case of Flix and coworking spaces in the case of Riba-Roja) made a great difference. In some cases, even people from Flix prefer to go to work in Riba-Roja:

> In Flix's coworking office, which is inside a business incubator, there is space but there is not a [*coworking*] philosophy. There is no dynamising figure, it is a different profile. [...] The coworking philosophy is very important. If people don't have a coworking philosophy, it doesn't make much sense. There is a whole idea behind the sharing economy, about living differently, in which we all agree.
>
> (Coworker of Zona Líquida, Riba-Roja d'Ebre)

7.6 Discussion and conclusion

This research underlines differences between urban and rural coworking spaces concerning one of the main characteristics of coworking: the community (Garrett, Spreitzer, and Bacevice, 2017; Spinuzzi et al., 2019; Rus and Orel, 2015). This chapter decomposes the notion of community by studying the different types of ties depending on the geographic distance and the intensity of the interactions. Different levels are considered: the level of the space, the immediate environment (municipality) and, taking a larger focus, the county and the regional levels.

The research results show that one of the main challenges of the development of rural coworking is to overcome the gap between an external and an internal community. The experiences of Cowocat Rural show that in some cases, opening a physical space at an early stage can be counterproductive, as the resources are used on material assets instead of community-building. These cases often fail as spaces remain underused and incur high fixed costs. The best practices identified by Cowocat Rural network consist on simultaneously developing a collaborative space and a community around it, in a progressive manner as represented in Figure 7.1. For example, some cases succeeded by identifying an initial core community, which later expanded once a physical space was adapted to help the visibility of the project. In some other cases, adapting a temporary physical space served as an exploratory test to attract people potentially interested in coworking. The results obtained at each stage of development provide feedback to plan the following steps, either putting more effort in constituting a community or in further concretising the project of opening a permanent physical space dedicated to coworking. This step-by-step strategy allows the optimisation of dedicated resources and avoids the temptation (common among municipal decisionmakers) to open a space at an early stage. It also prioritises the identification of potential coworkers and the progressive construction of the community, while adapting the project to the

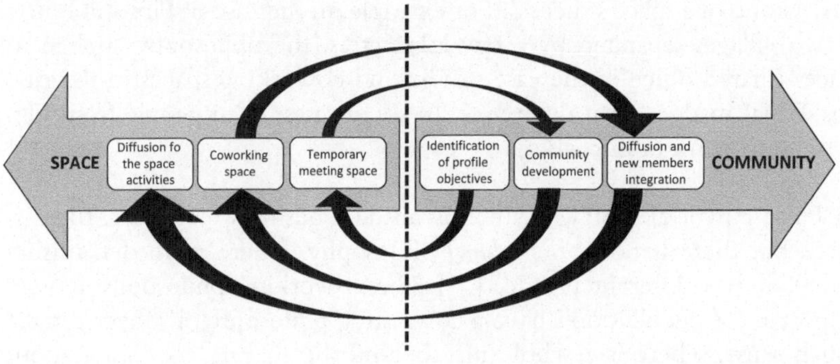

Figure 7.1 Virtuous circle of community development in a rural coworking space.

current needs and degree of evolution of the core community and the wider external community.

The results also underline the importance of the embeddedness of the coworking practices in the local environment in the case of rural coworking. One of the reasons is the difficulty of reaching a critical mass of members caused by low population densities. In rural areas, distances and transport time mean that interested individuals cannot participate in the co-located activities on a regular basis, but desire to be in frequent contact with other independent workers. Another reason justifying the local embeddedness of coworking is that the regional and local policies aimed at economic and social development include coworking as a mechanism of social integration and economic development (Avdikos and Merkel, 2020).

In the case of urban coworking, the focus has been on the internal collaborative dynamics, both from the academic and the policymaking perspectives (Jakonen et al., 2017; Blagoev, Costas, and Kärreman, 2019). Internal dynamics around specific domains and shared interests are more common in urban environments, as density allows the development of more specialised communities. In urban settings, collaboration among members can be more fruitful than with external local agents (Leclercq, Vandelannoitte, and Isaac, 2016; Gandini, 2015). In contrast, in rural coworking, professional collaboration among coworkers is often facilitated more by the convenience of co-location than by the complementarity of profiles.

Nevertheless, and in a more important degree than in urban settings, rural coworking created strong links with external agents. The analysis by levels of community reflects the importance that coworking promoters and managers in rural contexts attach to the impact on the territory, not only in the space itself. First, coworking contributed to social cohesion (Avdikos and Merkel, 2020). Working in public spaces, open to the public, allowed members to socialise and gain visibility among their neighbours. These new connections also resulted in local business opportunities in certain cases, that would rarely happen if freelancers had stayed working from home (Fuzi, 2015). Second, it created relational links in the territory by developing local professional networks. The existence of a stable physical space where to meet facilitated the organisation of events (Akhavan et al., 2019), initially often to diffuse the concept of coworking, but progressively to create business relationships (Capdevila, 2021). Third, coworking serves as a tool for municipal and regional government to design proposals to counter a rural exodus and offer new services to local freelancers. In many of the studied cases, the projects were publicly funded and promoted, like any other public service for the municipality (Avdikos and Iliopoulou, 2019). In that respect, spaces can be considered resilient spaces (Gandini and Cossu, 2019; Avdikos and Iliopoulou, 2019) as the initiatives pursue a social goal and the sustainability of a local economic model (Fuzi, 2015; Jamal, 2018), very far from real estate business models in urban spaces. Even though it has not yet proven successful, coworking has also been considered by village councils as a way to attract new talent (considering the potential post-pandemic urban exodus), and offer solutions to digital nomads (Clifton, Füzi, and Loudon, 2019).

References

Adler, Paul S., and Charles Heckscher. 2006. 'Towards Collaborative Community'. In: *The Firm as a Collaborative Community: Reconstructing Trust in the Knowledge Economy*, edited by Charles Heckscher and Paul S. Adler. New York: Oxford University Press. https://doi.org/10.1023/A:1021258713850.

Akhavan, Mina, Ilaria Mariotti, Lisa Astolfi, and Annapaola Canevari. 2019. 'Coworking Spaces and New Social Relations: A Focus on the Social Streets in Italy'. *Urban Science* 3(2): 1–11. https://doi.org/10/ggvfcj.

Avdikos, Vasilis, and Eirini Iliopoulou. 2019. 'Community-Led Coworking Spaces: From Co-location to Collaboration and Collectivization'. In: *Creative Hubs in Question*, edited by Rosalind Gill, Andy C. Pratt, and Tarek E. Virani, 111–129. Cham: Springer International Publishing. https://doi.org/10.1007/978-3-030-10653-9_6.

Avdikos, Vasilis, and Janet Merkel. 2020. 'Supporting Open, Shared and Collaborative Workspaces and Hubs: Recent Transformations and Policy Implications'. *Urban Research and Practice* 13(3): 348–357. https://doi.org/10/ggvfdh.

Blagoev, Blagoy, Jana Costas, and Dan Kärreman. 2019. '"We Are All Herd Animals": Community and Organizationality in Coworking Spaces'. *Organization* 26(6): 894–916. https://doi.org/10.1177/1350508418821008.

Brown, Julie 2017. 'Curating the 'Third Place'? Coworking and the Mediation of Creativity'. *Geoforum* 82: 112–126. https://doi.org/10/gbf8nk.

Butcher, Tim. 2018. 'Learning Everyday Entrepreneurial Practices through Coworking'. *Management Learning* 49(3): 327–345. https://doi.org/10/gf37h8.

Cabral, Victor, and Willem van Winden. 2016. 'Coworking: An Analysis of Coworking Strategies for Interaction and Innovation'. *International Journal of Knowledge-Based Development* 7(4): 357–377. https://doi.org/10/ggvd8n.

Capdevila, Ignasi 2015. 'Co-Working Spaces and the Localised Dynamics of Innovation in Barcelona'. *International Journal of Innovation Management* 19(3): 1540004. https://doi.org/10/gd3wg8.

Capdevila, Ignasi 2021. 'Spatial Processes of Translation and How Coworking Diffused from Urban to Rural Environments'. In: *Culture, Creativity and Economy*, edited by Brian

J. Hracs, Taylor Brydges, Tina Haisch, Atle Hauge, Johan Jansson, and Jenny Sjöholm, 1st ed., 95–108. London: Routledge. https://doi.org/10.4324/9781003197065-8.
Clifton, Nick, Anita Füzi, and Gareth Loudon. 2019. 'Coworking in the Digital Economy: Context, Motivations, and Outcomes'. *Futures*, July: 102439. https://doi.org/10/ggvff6.
Coll-Martínez, Eva, and Carles Méndez-Ortega. 2020. 'Agglomeration and Coagglomeration of CoWorking Spaces and Creative Industries in the City'. *European Planning Studies*, November: 1–22. https://doi.org/10.1080/09654313.2020.1847256.
de Peuter, Greig, Nicole S. Cohen, and Francesca Saraco. 2017. 'The Ambivalence of Coworking: On the Politics of an Emerging Work Practice'. *European Journal of Cultural Studies* 20(6): 687–706. https://doi.org/10.1177/1367549417732997.
Fasshauer, Ingrid, and Cathy Zadra-Veil. 2016. 'Quel Entrepreneur Pour le Coworking en Milieu Rural ?' *Entreprendre et Innover*2016(4): 17–24.
Flipo, Aurore. 2020. 'Espaces de Coworking et Tiers-Lieux: Les Réseaux d'une Nouvelle Ruralité ?' *Études Rurales* 206(December): 154–174. https://doi.org/10.4000/etudesrurales.23887.
Fuzi, Anita. 2015. 'Co-Working Spaces for Promoting Entrepreneurship in Sparse Regions: The Case of South Wales'. *Regional Studies, Regional Science* 2(1): 462–469. https://doi.org/10/ggvfb7.
Gandini, Alessandro. 2015. 'The Rise of Coworking Spaces : A Literature Review'. *Ephemera: Theory and Politics in Organization* 15(1): 193–205.
Gandini, Alessandro. 2016. 'Coworking: The Freelance Mode of Organisation?' In: *The Reputation Economy. Understanding Knowledge Work in a Digital Society*, 97–105. London: Palgrave Macmillan UK. https://link.springer.com/chapter/10.1057/978-1-137-56107-7_7#citeas.
Gandini, Alessandro, and Alberto Cossu. 2019. 'The Third Wave of Coworking: "Neo-Corporate" Model versus "Resilient" Practice'. *European Journal of Cultural Studies*, December, 136754941988606. https://doi.org/10/ggvfgf.
Garrett, Lyndon E., Gretchen M. Spreitzer, and Peter A. Bacevice. 2017. 'Co-Constructing a Sense of Community at Work: The Emergence of Community in Coworking Spaces'. *Organization Studies* 38(6): 821–842. https://doi.org/10/ggvd8z.
Institut Cerdà/AMB. 2019. *Els Espais de Coworking a les Ciutats.». Informe d'aprofundiment De l'economia Metropolitana. 17*. Barcelona: Institut Cerdà/Àrea de Desenvolupament Social i Econòmic de l'AMB. amb.cat/es/web/desenvolupamentsocioeconomic/actualitat/publicacions/detall/-/publicacio/els-espais-de-coworking-a-lesciutats/7982091/11708.
Jakonen, Mikko, Nina Kivinen, Perttu Salovaara, and Piia Hirkman. 2017. 'Towards an Economy of Encounters? A Critical Study of Affectual Assemblages in Coworking'. *Scandinavian Journal of Management* 33(4): 235–242. https://doi.org/10/gfzszr.
Jamal, Audrey C. 2018. 'Coworking Spaces in Mid-sized Cities: A Partner in Downtown Economic Development'. *Environment and Planning a: Economy and Space* 50(4): 773–788. https://doi.org/10/gdwjct.
Leclercq-Vandelannoitte, Aurélie, and Henri Isaac. 2016. 'The New Office: How Coworking Changes the Work Concept'. *Journal of Business Strategy* 37(6): 3–9. https://doi.org/10/ggvd7p.
Liimatainen, Karoliina. 2015 *Supporting Inter-Organizational Collaboration in Coworking Clusters: The Role of Place, Community and Coordination*. Espoo, Finland: Aalto University.
Mariotti, Ilaria, and Mina Akhavan. 2020. 'Exploring Proximities in Coworking Spaces: Evidence from Italy'. *European Spatial Research and Policy* 27(1): 37–52. https://doi.org/10.18778/1231-1952.27.1.02.

Mérindol, Valérie, Nicolas Aubouin, and Ignasi Capdevila. 2021. 'Combiner Confiance Résiliente et Réflexive, Hiérarchie Formelle et Prix au Sein des Communautés : Le Cas des Open Labs'. *Management International* 25: 184–205.

Merkel, Janet 2019. '"Freelance Isn't Free." Coworking as a Critical Urban Practice to Cope with Informality in Creative Labour Markets'. *Urban Studies* 56(3): 526–547. https://doi.org/10/gfkbg8.

Micek, Grzegorz. 2020. 'Studies of Proximity in Coworking Spaces: The Basic Conceptual Challenges'. *European Spatial Research and Policy* 27(1): 9–35. https://doi.org/10.18778/1231-1952.27.1.01.

Parrino, Lucia. 2015. 'Coworking: Assessing the Role of Proximity in Knowledge Exchange'. *Knowledge Management Research and Practice* 13(3): 261–271. https://doi.org/10.1057/kmrp.2013.47.

Rus, Andrej, and Marko Orel. 2015. 'Coworking: A Community of Work'. *Teorija in Praksa* 52(6): 1017–1038.

Spinuzzi, Clay 2012. 'Working Alone Together: Coworking as Emergent Collaborative Activity'. *Journal of Business and Technical Communication* 26(4): 399–441. https://doi.org/10.1177/1050651912444070.

Spinuzzi, Clay, Zlatko Bodrožić, Giuseppe Scaratti, and Silvia Ivaldi. 2019. '"Coworking Is About Community": But What Is "Community" in Coworking?' *Journal of Business and Technical Communication* 33(2): 112–140. https://doi.org/10/ggvd9c.

Part 3

Open labs at the origin of new governance models for innovation

8 Cracking the nut from the inside

Innovating from the ground up in highly constrained systems

Olivier Irrmann

Research on innovation tends to predominantly examine cases of organisations from the private sector, evolving in competitive markets, often with a fast pace of change. However, a large part of our life in society is orchestrated by non-competitive forces, with institutions and structures that tend to change rather slowly but are still able to innovate. Two such forces are national states or local authorities that design and provide public services, and the education sector, which in most countries is structured outside of competitive markets. These two domains are highly regulated, aim at providing stability and reproducibility, and promote equal access to services to citizens. Innovation is possible but does not follow the traditional paths that are analysed in the private sector, and therefore many classical recommendations might not be applicable or at least require considerable modifications. In this chapter we analyse the evolution of two organisations that can be considered as open labs and that played an important innovative role in these highly constrained sectors (HCS) in France. We try to understand how and under which conditions they had been able to become major triggers of innovation in sectors that are not as conducive to fast-paced change as some of their private counterparts.

The Centre de Recherches Interdisciplinaires (CRI) and La 27è Région are two instances of open labs that have had particularly successful trajectories, working at the interstices between organisations and professional domains, led by a strong vision of the public good, and creating links between creative individuals and strongly regulated and constrained institutions. Both organisations are involved in exploration and experimentation and do not have a catalogue of services, rather providing opportunities to test, run, and deploy a new way to operate. Though physical spaces have been important in their development, it is far from being central in their operations, and the availability of a place was not a prerequisite for starting their operations.

The first organisation is the CRI, a not-for-profit organisation based in Paris and a pioneering innovative education and research organisation. It has become a reference in France for championing interdisciplinarity, deploying new educational programmes and creating a culture of openness and collaboration between educational institutions. The CRI was cofounded by two researchers in Biology, Francois Taddei and Ariel Lindner. The first projects

DOI: 10.4324/9781003125587-12

of the CRI started informally in 2003 and the official not-for-profit structure (Association Loi de 1901) was created in 2006. Its founders have been involved in several governmental reports about the future of education, and the CRI has now become one of the important and impactful voices about a new way to think about education and learning in France. In 2020, the CRI was employing around 50 people and had an annual budget nearing €10 million.

The second organisation is La 27è Région, a not-for-profit organisation also based in Paris that defines itself as a laboratory for public transformation. It insists on the fact that it is not a consulting firm, nor a think tank or a research centre, but a "do-tank" that helps public institutions rethink their approach to the design of public policy and public services. It combines its deep expertise of the mechanisms of public administration with a long practice of participative and user-centred design, and applies it to transformation projects with local and regional authorities. La 27è Région is one of the most well-known actors in public policy transformation in France. Its co-founders are Christian Paul, an elected politician, Stéphane Vincent, at the time a consultant specialised in public administration and economic development, and Daniel Kaplan, an activist and entrepreneur in the world of digital services. It started in 2008 as a project incubated within the Fondation Internet Nouvelle Génération (fing .org) in cooperation with an association of regional authorities. It became an independent organisation in 2012, also taking the form of the "Association Loi de 1901", currently operates with an annual budget of €1 million, and employs a team of ten people with backgrounds in management, political science, design, social innovation, and social sciences. Stéphane Vincent is still leading the organisation.

Data about the cases was collected over a period of several years (2015–2021), through recorded and transcribed interviews with founders, managers and users of the open labs, secondary data, participant observation at the CRI over a period of several months, and participation to the activities of the board of La 27è Région. The author has also been involved in partnerships with both organisations for the design of training programmes and research projects, bringing unique insights into their operational functioning and their organisational culture.

In the next sections we briefly present the specificities of the highly constrained systems in which CRI and La 27è Région operate, we examine how these open labs progressively impacted their respective sector through the process of building a strong "middleground", and finally what role physical spaces have played and will play in their operations.

8.1 The constraints of managing transformation in the French public administration and education sector

The two organisations we are considering in this chapter are involved in the transformation of the public administration and the educational system. These two sectors display a high level of centralisation, with a top-down decision-making logic and a very strong hierarchical culture that make it difficult to conduct reforms or implement change. These systems are constructed with a logic of reproducibility at a very large scale, which in this case is at the national level.

Public administration and education in France are important sectors with a large number of employees. There are more than a million employees at the level of primary and secondary education, over 90,000 educators at university level, and close to 1.9 million territorial civil servants (fonction publique territoriale). Most of the people working in these two sectors are civil servants, mainly recruited through competitions (concours). As a result, these national systems tend to produce organisational routines and rules that are going to be applied to several hundreds of thousands of employees and will impact an even larger large number of users. The order of the day for these sectors is not to produce tailored and fine-tuned solutions and services, but to have reproductible systems at the national level with limited variability.

National systems are challenging to modify, as they address and confront cultural systems of representations built over several decades, but also because any modifications will have an impact on a very large number of actors across the administration, including citizens and beneficiaries of public services. Cole (2010) noted that attempts to reform the French state face strong institutional traditions with an administration

> built upon a hierarchy within the state, upon a meritocratic form of recruitment, upon an internal career structure based mainly on seniority, on an ethic of public service and a belief … that the State embodies the general will and speaks in the name of the general interest of the nation as whole.

Even though decentralisation in France has progressed a lot since the 1980s, central government is still involved in the delivery of many public services. Even the coordinated efforts for transforming and modernising the French state since 2008 are themselves structured within a dedicated central administration, under the name of "Direction Interministérielle de la Transformation Publique" (DITP).

The French education system is notoriously difficult to reform. Analysing the 1950–1986 period, Weiler (1988) noted that an enormous proliferation of reform proposals had peacefully coexisted with an extremely limited degree of actual change in the education system. The low level of implementation can be traced back to a tension between the traditional vision of education as a way to promote equality and the demand for greater excellence and the production of a technical and productive elite. This paradoxical tension between a vision of republican equality and the creation of a "school nobility" (Iribarne, 2008) can be observed in the structural dichotomy in higher education between the "grandes écoles" and the universities and the primacy of theoretical teaching in the school system, leading to a system that promotes free education but also focuses on training students for examinations and competitions (Allouch, 2017).

All these features tend to reduce the importance of collaborative processes in the vision of education. The French education system has experienced an important movement of decentralisation, with many decisions delegated to local education authorities (up to 50% in some estimates), even though centralisation is still important, such as for the definition of the national curriculum, the certification of teachers and lecturers, and the issuance of national degrees (Zanten and Robert, 2000).

In the highly centralised and constrained systems of education and research, the logic of submission and hierarchy does not necessarily promote individual talents or the emergence of new objects, programmes, and protocols. Most activity (at least in education) focuses on the reproduction of set programmes, with a very centralised management of careers. There is a very limited element of subsidiarity in the deployment of careers, for instance schools do not recruit their own teachers independently, they are nominated by a central authority.

In terms of public policy formation, the practices tend also to be centralised with very limited participation from beneficiaries. There is a general tendency to consider that specialists appointed by the administration can perfectly understand the needs of the citizen and therefore can design policies and rules for them, not with them. The design of public policies therefore considerably lags behind the practices of technological and service industries that have embedded user-centeredness in their practices for decades.

Within these highly constrained environments, the CRI and La 27è Région have successfully conducted creative experiments and have been important drivers of innovation. In the case of the CRI, a community of educators and researchers set up over a decade a powerful middleground for innovation. Starting from a single table in a corridor at the university, they progressively hacked their way from inside the education system, introduced numerous pedagogical innovations and became one of the main representatives of interdisciplinarity in the educational system. In the case of La 27è Région, its founders introduced design approaches in public policy and created spaces for creativity and deep knowledge of users within the public administration, including a space for experimentation, leading to the emergence of considerable innovation in the administration and public service. The two cases illustrate how even in extremely constrained, centralised, and hierarchical fields, innovation and disruption can be negotiated and progressively percolated through the system.

8.2 Open labs as the building blocks of a middleground in HCS: from activism to the construction of a community of innovation

The CRI and La 27è Région form two instances of open labs that could be described as constitutive of a strong middleground (Cohendet et al., 2010, 2014; Grandadam et al., 2012). The notion of middleground emerged in studies of the "anatomy of creative cities" (Cohendet et al., 2010), trying to understand why creative processes emerge and remain in specific local ecologies of knowledge. Cohendet et al. (2010) propose three layers that form the basis of creative processes: the underground, the middleground, and the upperground. The upperground is composed of innovative firms and institutions that focus on the integration of dispersed knowledge and are able to finance and test new forms of creativity on the market. The upperground can be formed by firms from different sectors, research institutes, universities, or artistic centres. In our case, the upperground is typically formed by central and local administrations, ministries, local authorities such as cities and boroughs, universities, and national research organisations.

The underground "brings together the creative, artistic and cultural activities taking place outside any formal organization or institution based on production, exploitation or diffusion" (Cohendet et al., 2010). This layer is mainly composed of individuals or small collectives that share a deep interest for their creative activities – up to the point of shaping their identities and their lives – but wanting to remain outside the institutional logic of exploitation and focusing on exploration. In the HCS we examine here, the underground is made of teachers, educators, researchers, civil servants, activists, students, social entrepreneurs, or citizens interested by the development of public services or the evolution of the education system.

The middleground is an intermediate structure linking the underground to the upperground. It navigates between the formal and informal world, playing a brokerage role to promote both exploration and exploitation mechanisms around creative ideas. The fabric of a middleground is in the groups and communities that will link the informal and mostly individual underground culture with formal organisations and institutions of the upperground. The concept of the underground–middleground–upperground is a powerful instrument for understanding the dynamics of innovation in industrial or creative sectors (Cohendet et al., 2014; Sarazin et al., 2017). Middlegrounds do prosper when they are structured around a set of places (the physical and local anchor of creative activities), spaces (the cognitive dimension of knowledge creation and codification), events and projects that provide a structural canvas to the development of ideas, and the generation and diffusion of knowledge (Grandadam et al., 2012). The strength of a middleground is instrumental for the emergence, development, and resilience of an ecosystem of innovation. Cohendet et al. (2020) showed how the evolution of the innovative ecosystem in the video game industry in Montréal has been mediated by the existence of a rich middleground, allowing very different actors in the industry to interact, learn from each other, develop transversal projects, promote local talents, and entrepreneurial initiatives.

The nature of HCS does not favour the existence of intermediate structures that allow or promote transversal collaboration. The structure of the French public administration and the education sector is highly centralised, which promotes a culture of top-down decision making and a limited possibility for people within the system to influence decisions. It is therefore important to create opportunities for such collaboration and to orchestrate activities that are inherently difficult to conduct in HCS. Open labs can play an important role in triggering the emergence and the development of a middleground in HCS.

In our two cases, we can observe a similar pattern, starting with the structuration of an epistemic community dedicated to the advancement of a specific cause (education through research for the CRI and the conception of public policies through design approaches for La 27è Région). This community will launch the first experiments and projects, often in an informal fashion, collect data and structure knowledge about their approaches to inform deciders and involve members of the underground. When the experiments reach a critical mass, the community will officially structure its action through the creation of an open lab that will solidify the knowledge dissemination process, bring an official face to events, and create a legal entity that will have the ability to be an official partner for consortia and acquire funding and resources. With this combination of a strong epistemic community backed by the construction of a legal entity under the form of an open lab, the middleground can develop further and fully play its role of bridging groups of innovative individuals to the institutional level that can allow the scaling-up of experiments. The diffusion of knowledge and ideas through publications, reports, blogs, white papers, conferences, courses, and training sessions will solidify the "cognitive space" dimension of the middleground, making the open lab an important part of meaningful discussion about the issues the community is interested in.

In the two cases studied, the story started with a group of people determined to innovate and change practices within highly constrained sectors. Beyond being simply leaders with a strong legitimacy in their respective fields, they formed an epistemic community, "a network of professionals with recognised expertise and competence in a particular domain and an authoritative claim to policy-relevant knowledge" (Haas, 1989, 1992;Cohendet et al., 2014). The goal of an epistemic community is to generate, share and curate knowledge to have an impact on society. In our cases the epistemic communities at the origin of the open labs have leveraged physical spaces for talking about and testing innovation, then progressively orchestrated resources to get exposure and recognition, and then gained leverage to influence the system through special projects and exemptions to the rule. As the central actors in the community are recognised professionals, they can leverage some legitimacy when starting projects or making proposals. Both of our projects started with a small group of very motivated founders, people embedded for many years into the core activities of the HCS and trying to address their own feeling of inadequacy of the system and channel their willingness to create change.

The CRI started as a project led by researchers in biology, addressing the need for more interdisciplinarity in life sciences and wanting to experiment in new ways of teaching and learning. The two co-founders of the CRI, François Taddei and Ariel Lindner, had gained considerable experience within interdisciplinary research groups, with a very high level of creativity and the ability to conduct experiments at the edge of scientific inquiry. These work environments nurtured a taste for experimentation and a "just do it" approach that we find later in the texts and calls to action that shaped the CRI.

Under the coordination of Taddei and Lindner, a group of researchers started organising an interdisciplinary seminar in 2003 that was mixing researchers and students. It got successful enough to trigger demand for the creation of a Master 2 programme in 2004, followed by a special programme, "Science Academie", involving high school students, then a doctoral programme, a Master 1, and finally a Bachelor programme. Between 2003 and 2009, what started as an informal meeting in a corridor had become a full-fledged programme from Bachelor's to PhD based on the values of learning through research and experimentation (Taddei, 2013). In 2006 the CRI had gathered enough attention and the willingness of major funders to finance its operations and projects that it required the creation of a legal entity. Taddei and Lindner created an official vehicle for the CRI under the form of a non-profit organisation (Association Loi de 1901). It marks the birth of what we can describe as an open lab dedicated to new forms of education.

The CRI, often through the media exposure of its co-founder François Taddei, is a prominent voice defending a participatory approach of education, letting the voice of students rise, and promoting agency from children and young people. What separates the CRI from other advocates of a different pedagogy is its ability to quickly experiment and launch projects and programmes. This has been largely facilitated by the financial contribution of the Bettencourt-Schueller Foundation that has supported the CRI for most of the past 15 years. Among the impressive achievements of the CRI is the creation of the first interdisciplinary doctoral school in France. A doctoral candidate can apply only if he/she can prove that the topic cannot be possibly addressed by a single domain of science within a classical monodisciplinary doctoral school. The creation of this doctoral programme required a change in the law that was negotiated with authorities and the Ministry of Science and Research. Another important programme developed at the CRI is "les Savanturiers" (a play on words from savants (learned person, scholar) and aventuriers (adventurers)) that promotes a pedagogy of research in secondary and primary schools by transforming pupils into researchers and acculturating them to the research process. Les Savanturiers has now trained over 30,000 teachers, conducted projects among 30,000 pupils, and has a network of 400 researcher mentors that are coaching classes. A similar path has been taken since 2019 by the project "Teachers-Researchers", a collegial citizen science framework "with the intention to create a giant research laboratory in which educators produce open, trustworthy, transferable, and ever-evolving practice-based evidence on how to address the many challenges of education". Most CRI projects have a connectivity dimension ingrained into them which is central in a middle-ground, but they also focus on the pursuit of knowledge generation and sharing that is the mark of epistemic communities.

For La 27è Région we have founders with a long experience of working for regional authorities with a specific focus on digitalisation and internet services for citizens. Stéphane Vincent worked for a few years in a Regional Development Agency in Limousin, in the centre of France, then created a consulting company before launching the La 27è Région project. Christian Paul, co-founder and first president of the association, was an elected member of parliament for over 20 years (1997–2017) with close ties with the different ministries in charge of civil service and state reforms. La 27è Région started in 2008 as a project incubated in another organisation. At a time when France was divided into 26 regions (the number was reduced to 18 after the 2016 mergers), the goal was to find a new type of governance for regions in France, thus creating the vision of an ideal 27th region. The goal was to provide a tool to regional authorities for rethinking the design of public policy. It took four years of activity before the 27th region became an independent organisation in 2012.

They developed a series of extremely innovative protocols, the most well-known being "Territoires en residence" and "La Transfo". The "Territoires en résidence" programme completely immerses a multidisciplinary team in a public facility or service – a neighbourhood organisation, school, community centre – for three separate one-week periods to question the activities of the public service provider from the standpoint of its beneficiaries in order to propose concrete improvements. Nineteen experiments were conducted between 2009 and 2012. The "La Transfo" programme started in 2011 and seeks to prototype an innovation "lab" within a public organisation. A multidisciplinary team of residents takes up shop within the organisation department for a total of seven to ten weeks over a period of one to two years to play the role of temporary laboratory, working with the public agents, elected officials, and citizens.

La 27è Région leverages a network of 120 partners and colleagues that are associated with its projects, bringing *ad hoc* expertise in sociology, anthropology, design, architecture, videography, or arts. It works exclusively with public collectivities, such as regions, cities, departments, or ministries. In the 13 years since its creation, La 27è Région has become a catalyst of new practices in the system of innovation associated with public policymaking. There is hardly an event where it does not have a representative and will bring a testimony about new practices for connecting with citizens, experimenting solutions, and helping collectivities to redesign their approach for conceiving public services. It is recognised as one of the major interfaces between design practices and the reform of the state, playing an important role in the shift from neo-managerial practices inherited from consulting firms to an attention to the needs of users and citizens (Alauzen and Malivel, 2020). La 27è Région has facilitated the progressive emergence of innovation laboratories in public administration, even though the diffusion of user-centred and design practices is still at an early stage (Coblence and Vivant, 2017; Irrmann, 2020). A remarkable trait of La 27è Région is its capacity to constantly produce documentation of all its protocols, experiments, and projects. Every project is thoroughly described, with a review of the results, blog posts, and user manuals. The goal is not to keep this know-how in-house but to disseminate as much as possible, even to non-members of the association. When the team of La 27è Région feels that there is nothing more to be learned, they generally stop the projects in order to focus on the next meaningful challenges. This shows the strength of the epistemic community culture that support the activities of the open lab.

8.3 The role of physical and digital places managed by open labs in the development of middlegrounds

In our two open labs, the importance of a place – the physical dimension of the middleground – is rather variable. Creative and innovative communities often need places to convene and further develop their feeling of belonging, but these spaces can remain quite modest for a long period of time. The CRI started by occupying a few tables in a corridor. La 27è Région started by using a desk within another association. These spaces, when they exist, do not need to be permanent and they will often be symbolic. They could be a meeting in a corridor or in a canteen after service hours, or in a coworking space or a café at regular intervals. Both our case organisations have developed places over the years and followed a different path. The challenges of dealing with real estate are complex, costly, and often demand a long time before they are solved. Open labs do with the space what they can, as different spatial affordances will allow different things.

For La 27è Région, the opening of "Superpublic" in 2014 – a 300 m^2 space in the 11th Arrondissement in Paris – was the opportunity to have its own place for organising workshops and seminars. The space was co-occupied with a couple of external design studios and sometimes staff from partner public collectivities, which allowed a high level of collaboration. A few years later, La 27è Région got involved with the "Halles Civiques" project, a large building of 650 m^2 in the 20th Arrondissement of Paris, a coworking space shared with several small organisations mainly focusing on social innovation. The Halles Civiques enabled communication with social innovation activists and their projects. La 27è Région took the lead in organising the Halles Civiques, as a tenant and a manager of tenants The 2020–2021 Covid lockdown put a halt to the activities of the Halles Civiques, drying up the stream of revenue. The Halles Civiques (Association Loi de 1901) ceased operation and was liquidated in 2020. Starting 2021, La 27è Région had left its historical buildings and rented spaces in another coworking space, "Oasis21".

For CRI, there were humble beginnings in the corridors at the university, it then moved to a coffee room. Later, the group got a proper spot by occupying half of a floor at the Medical School at the University of Paris Descartes, but was spread between several locations, the research groups remaining within the laboratories of other teams. Part of the CRI stayed for some years in a 5,000 m^2 unrenovated former English language department building in the Marais, provided by the City of Paris. An important change occurred in 2016, when forced out of their Marais headquarters, the CRI started to look for a new space and found the opportunity to rent two floors of the Tour Montparnasse in Paris, the tallest building in the city. The Tour Montparnasse was going to be completely emptied within a few years for a major renovation, which created an opportunity for renting at favourable conditions. The location at the Montparnasse tower allowed the integration of other partners within their premises (among others the training collective CoDesign It and the online publication The Conversation France) and played an important "wow" effect with visitors, students, and any partners visiting their premises. In 2018, the CRI left the Montparnasse Tower and entered a newly renovated 7,000 m^2 building in the Marais (the one they had occupied a few years before), the result of an agreement between the CRI, the City of Paris which provided the building, and the Bettencourt-Schueller Foundation that funded most of the renovation. The new building has allowed the CRI to bring laboratories to the same premises as other educational activities for the first time in its history.

However, in the absence of the need for specific equipment, places by themselves do not play the most important role in the diffusion of new ideas. The real issue is that the most important places are somewhere else, in the locations where the practices of the middlegrounds are diffused and not in the spaces in which new practices have emerged. Though there is certainly an element of comfort and the possibility of intense interactions by controlling its own physical space, the activities of these two middlegrounds did not decline when access to spaces was limited during the lockdown of the years 2020–2021. They kept on having a strong influence and developing new projects and events.

Much like Hacking Health (see Chapter 5), the important places are the ones where events are organised, where experiments are conducted. For instance, the important place for the Savanturiers project (at CRI) are the thousands of classrooms where the pupil-researcher protocols are deployed. For La 27è Région it is the local collectivities and the myriad of administrations that seize and deploy their programmes and methods.

The digital places will also play an important role in the development of middlegrounds. There has been a recent boom in the organisation of virtual events, following the health constraints of 2020–2021 and many events have become much more accessible and much more ubiquitous. La 27è Région has started to regularly organise webinars for its members, and it radically increased the ability of its members to participate beyond a limited geographical area. The reach of these events and knowledge have greatly increased and therefore the impact of its middleground. At the CRI, digitalisation had already been quite advanced, and some projects greatly benefited from the expansion of access. Some projects like "Teachers Researchers" had actually developed at a much-increased pace following the lockdown. Physical presence in a place is a powerful vector of interaction and social contacts, but we might have underestimated the constraints of such a mode of interaction. In HCS, the ability to get away from your workplace is rather complicated, as one needs to get a mission order that must be negotiated with a whole array of administrative powers.

The ability to go to a digital place is actually probably the most powerful means of influence in HCS.

Platforms and digital tools will also play an important role to help navigate and diffuse knowledge. These issues have been present in large organisations for ages, as we saw with the long tradition of research on knowledge management. Some middlegrounds and open labs need to think about the kind of platforms they will use for diffusing their ideas. La 27è Région is exemplary for that, following a lower-tech approach based on the constant production of written and visual documentation on its website: dossiers, blog posts, reports. The CRI has a much less readable website but is developing a set of digital tools for its members and communities, such as the "Projects Platform" or developing new ones like "Open Pantheon" or the development of platform like "the GPS of education" powered by new algorithms and AI. We see here the double phenomenon of expansion and reduction of organisational space with ICT described by Mukherjee (2017).

8.4 The organisational conditions of success of open labs in HCS: interstitiality and independence

Looking at the evolution of La 27è Région and the CRI, we identify a few conditions that were at the heart of the success of their activities: first a condition of non-competition and "interstitiality", and second a condition of independence and "non-hijackability".

The two open labs show features of "interstitial organisations" (Bátora, 2013; Slaughter and Rhoades, 2004). For Batora (2013), an "interstitial organisation" is defined as

> an organization emerging in interstices between various organizational fields and contingent upon physical, informational, financial, legal and legitimacy resources stemming from organizations belonging to these different organizational fields. Also, interstitial organizations harbour various sets of conventional and alternative practice drawn from the various organizational fields they span and pertaining to the performance of their core functions.

To become a key player in the middleground, these open labs follow the principle of interstitiality, they grow up at the interstices between players of the ecosystem and facilitate the connection between domains of activity. The CRI has developed its programmes at the interstices between research areas and education institutions, where a field is ending, and the other is not starting yet. As such the CRI always provides a solution without threatening what exists, leveraging the goodwill of individuals within the partner institutions and hacking the system rather than threatening it or competing for scarce resources. La 27è Région similarly is always playing the role of facilitator between diverse actors and is always trying to conduct projects that have more than one "owner", mixing different levels of regional authorities. It provides skillsets that do not replace the local actors but provide the toolset and the coaching approach to let them conduct the diagnosis and the proposal. As in empathic design and codesign, they never design for, but design with. As a consequence, the organisations we studied were rarely seen as competitors in their institutional field.

The key challenge for these open labs is to remain in the area between underground and upperground, with the role of connecting, developing knowledge, and actionable innovation. The most strategic activity is the ability to develop the "spaces", that is – in terms of middleground theory parlance – the cognitive playground for ideas and knowledge about key topics, education, and research for the CRI and public transformation for La 27è Région. As in open-source communities, this space gets developed through symbolic recognitions on top of an extended culture of sharing. Nothing is hidden and therefore there is no temptation of resource capturing, all documentation is part of a contributive manifesto, and the organisational routines reflect this manifesto. The CRI has decided to diffuse a culture of a "research collaboratory". Information is shared, all activities must carry an element of interdisciplinarity (such as in the selection of PhD projects), experimentation and individual agency is found everywhere, many topics within the CRI research fellows community focus on citizens and open science. La 27è Région systematically documents all experiments and projects on its website, producing the cookbook of participative and citizen-centred public policy design.

The culture of contribution is an important element for impact in HCS. That is also an antidote against retortion measures from centralised institutions that do not want new players in their playground. The middleground is here to do things that are quasi-impossible to test in centralised organisations, leveraging its ability to experiment and use its creative slack. But in HCS it also requires organisational freedom and independent resources. One needs to remain conscious that shifting to a system of contribution is not that easy. It requires a certain discipline and the practice of sharing and diffusing. That is where private organisations could learn from these spaces.

Another important element that allowed our two cases to grow and conduct their operation as they saw fit was organising the conditions of independence. Both the CRI and La 27è Région used the powerful status of "Association Loi de 1901", a form of non-profit organisation that is used in France. For the CRI the status of non-profit was a way to get protected and still have mastery over their resources. The association had to be created when the Foundation Bettencourt Schueller started to provide ample funding in the order of several million euros to promote the interdisciplinary approach in life sciences. The non-profit structure provided a well-defined organizational form for conducting the project and the ability to use the funds without having to deal with the arcane accounting rules of a university. For La 27è Région, the status of non-profit is also a way to communicate the goals of the organisation, that will not be profit but a focus on deploying and testing ideas and concepts. There is a clear motto that "we do not seek contracts, we conduct experiments".

Its network of partners can benefit from its access to local collectivities and develop its consulting practice if it wants, with the limitation that all projects with La 27è Région need to be documented. The legal structure also allowed the independent use of a US$1.4 million subvention from the Bloomberg Foundation received in 2017.

Having an independent status is also a way to guarantee that the project will not be hijacked by the partner institutions, either for political reasons, power games, or to obtain financial resources. The hijacking was seen as an important threat that occurred regularly and would have mostly killed the transformative power of the organisation. Independence also came with the development of a collaborative and contributive form of governance. Both open labs have developed a set of principles and practices for escaping cronyism and possible manipulation.

La 27è Région is a smaller team but has a board of members that is rather extended. The members of the association are mainly "collectivités" (at €5,000/year) but a few years ago membership was also opened to individual members (at €40/year). The collectivities are sites of experimentation; they are the main beneficiaries of the knowledge and the main sites of the projects. Individual members are more the friends of the association, composed of researchers, designers, or politicians. The board of the association is composed of approximately 20 members who meet four times a year to exchange information about the different projects, and the annual general assembly attracted 60 participants at the last count in June 2021. There is no hidden agenda, or hidden issues. All problems, including financial and operational, are discussed openly among all the members.

The CRI also has different boards, including a board of directors and a scientific advisory board. Issues are generally shared in a culture of openness. The tradition is to rely mainly on the scientific advisory board (SAB) for decisions about research fellows' nominations. The SAB is composed of eminent researchers from around the world and carries the final decision about many issues, particularly those where power games could play out, like the recruitment and funding of young fellows. It is a rather elegant way to shun the temptations of pressures and revenge – "it's not our decision, it's the SAB's".

8.5 Conclusion

The challenge of innovating in HCS can be addressed by organisations that are focusing on building a strong middleground in their ecosystem. The middleground will foster links between the creative underground and the production-focused upperground. In HCS, such as public institutions or the education and research system, the upperground will often be composed of very large organisations and institutions, built on a strong system of regulations and impacting many hundreds of thousands of people with any change that occurs. The CRI and La 27è Région are in the role of interstitial organisations. These types of open labs are most likely to have a lasting impact on the constitution of a middleground in HCS. By their very nature, at the interstices of several domains, open labs as interstitial organisations do allow experimentation and provide an "acceptable space for deviance" to test ideas, deploy new solutions, and harness the creativity of people within these HCS. The nature of interstitial organisations is such that they are not competing with existing institutions, and do not threaten the established order while allowing the experimentation with new approaches. The not-for-profit French organisational form, the Association Loi de 1901, provides independence and the ability to decide rapidly without having to pass through the loops of the administrative bureaucracy. In both our cases they also allowed control over important financial resources that could be deployed as close as possible to the core activities.

Interstitial organisations are served by the creative communities they aggregate, many of them placed into the fabric of the system that needs to be changed. The goodwill of the participants and the deep knowledge of the realities of the field provide a powerful combination. The core engine of the organisation can remain quite small compared to the impact it has. For La 27è Région, six to ten people are employed by the structure. For the CRI it has grown over the years to 50 people, but even the co-founders are employed by other structures and are officially detached to serve the activities of the CRI. This structure is common in the French research environment, as many research laboratories are actually consortia composed of members coming from different institutions.

The strategic feature of the two open labs studied is their contributive dimension. It could be an important feature for thriving in HCS, where large and centralised institutions represent a dominating force and creative individuals have little agency beyond their very local environment. CRI and La 27è Région play an important role as engine of middlegrounds in their respective domains of activity. They set an example as being dedicated to exploration and dissemination of new practices and testbeds for innovations in education and the design of public services.

As such, they form a special case of open labs, and do not necessarily fit all the classical elements presented in Chapter 1. The spatial dimension and the importance of having a physical place are probably ones that might be revisited in some conditions. The CRI ended up having a rather consequent real estate footprint at its disposal, which grandly facilitates the testing of many initiatives. However, a large part of its activities will be deployed at other locations (Savanturiers), or in virtual spaces through ICT platforms (Teachers-Researchers, Learning Planet Festival). We also observed that La 27è Région developed quite a lot of its activities with the opening of real estate assets, starting with one location (Superpublic) and shifting to two different locations (Les Halles Civiques), leveraging the power of having a "territorial" anchoring, before the economic situation forced it to shut down both places. However, its activities didn't stop in any shape or form, showing that the core of La 27è Région's activity can be considered as place independent. The main asset here is not the place, but the ability to deploy activities and test innovative activities at the interstices between different institutions and organisations and play a structuring role as a middleground in a highly constrained ecosystem.

Our cases demonstrate that having a transformative impact is possible in a very constrained system, under certain conditions of independence, and interstitiality, in order to participate in the development of a middleground. This study shows that the notion of middleground can be applied beyond the creative industries and that it can play a role in understanding the dynamics of creativity, innovation, and change in the public sector or in strongly constrained and regulated systems. The key to a solid middleground is still the existence of an epistemic community that will transform the system from within. There is a strong dimension of activism in these environments but also a strong sense of a societal duty, the need to serve a higher good, and the necessity to contribute specific "commons"(as in Ostrom and Hess, 2007) , be it education or a good society with strong public services. What distinguished these two organisations is the ability to sustain an activity in the long term. Like all communities of innovation, time is important, and the mastery of middlegrounding is a skill that is long to develop.

References

Alauzen, M., & Malivel, C. (2020). Le design est-il en passe de devenir une science de gouvernement? Réflexion sur les espoirs suscités par les sciences du design dans la modernisation de l'État en France (2014–2019). *Sciences du Design*, 12(2), 36–47. https://doi.org/10.3917/sdd.012.0036.

Allouch, A. (2017). *La société du concours: L'empire des classements scolaires*. Paris: Seuil, Coll. La République des idées.

Bátora, J. (2013). The 'mitrailleuse effect': The EEAS as an interstitial organization and the dynamics of innovation in diplomacy. *JCMS: Journal of Common Market Studies*, 51(4), 598–613. https://doi.org/10.1111/jcms.12026.

Coblence, E., & Vivant, E. (2017). Le design est-il soluble dans l'administration ? Trois trajectoires d'institutionnalisation de l'innovation publique. *Sciences du Design*, 5(1), 52–68.

Cohendet, P., Grandadam, D., & Simon, L. (2010). The anatomy of the creative city. *Industry and Innovation*, 17(1), 91–111.

Cohendet, P., Grandadam, D., Simon, L., & Capdevila, I. (2014). Epistemic communities, localization and the dynamics of knowledge creation. *Journal of Economic Geography*, 14(5), 929–954. https://doi.org/10.1093/jeg/lbu018.

Cohendet, P., Simon, L., & Mehouachi, C. (2020). From business ecosystems to ecosystems of innovation: The case of the video game industry in Montréal. *Industry and Innovation*, 28(8), 1046–1076.

Cole, A. (2010). State reform in France: From public service to public management? *Perspectives on European Politics and Society*, 11(4), 343–357.

Grandadam, D., Cohendet, P., & Simon, L. (2012). Places, spaces and the dynamics of creativity: The video game industry in Montreal. *Regional Studies*, 47(10), 1701–1714.

Haas, P. M. (1989). Do regimes matter? Epistemic communities and Mediterranean pollution control. *International Organization*, 43(3), 377–403.

Haas, P. M. (1992). Epistemic communities and international policy coordination. *International Organization*, 46(1), 1–35.

Iribarne, P. d' (2008). *L'étrangeté française*. Paris: Éd. du Seuil.

Irrmann, O. (2020). Le design au service des territoires et des politiques publiques. *Horizons publics*, Hiver 2020, Hors-série. La 27e Région, http://la27eregion.fr.

Mukherjee, A. (2017). *Organizational space collapsed, organizational space expanded: Experiencing space with ICT, affordance and the body* [Doctoral dissertation]. Paris: Université Paris Dauphine.

Ostrom, E., & Hess, C. (2007). *Understanding knowledge as a commons: From theory to practice*. Cambridge, MA: MIT Press.

Sarazin, B., Cohendet, P., & Simon, L. (Eds.). (2017). *Les communautés d'innovation*. Caen, France: EMS Management & Société.

Slaughter, S., & Rhoades, G. (2004). *Academic capitalism and the new economy: Markets, state and higher education*. Baltimore, MD: Johns Hopkins University Press.

Taddei, F. (2013). Pour un enseignement interdisciplinaire. *Hermes, La Revue*, 67(3), 57–61.

Weiler, H. N. (1988). The politics of reform and nonreform in French education. *Comparative Education Review*, 32(3), 251–265.

Zanten, A. V., & Robert, A. (2000). "Plus ca change…"? Changes and continuities in education policy in France. *Journal of Education Policy*, 15(1), 1–4.

9 Living labs and innovation commons in healthcare ecosystems

The case of the TransMedTech Institute in Montréal

Nathalie Tremblay, Patrick Cohendet, Geneviève Cyr, Margaux Manent, Laurent Simon, Marie-Pierre Faure, and Carl-Éric Aubin

This chapter analyses the case of the TransMedTech Institute (iTMT) as a living lab in the healthcare ecosystem, highlighting how this organisation contributes to shaping the dynamics of the medical technology sector in Montréal, Canada, and abroad. iTMT was founded in 2016 from an initiative led by Polytechnique Montréal, an engineering university, with the collaboration of four other founding institutions, Centre Hospitalier Universitaire Sainte-Justine (CHUSJ), Université de Montréal (UdeM), Centre Hospitalier Universitaire de Montréal (CHUM) and the Jewish General Hospital (JGH). These five institutions came together to meet the needs identified by the medical and scientific community to support the development and integration of validated technological solutions, embedded in the clinical setting and in compliance with the highest quality standards of medical technology development.

iTMT could be seen as a core platform of an innovation ecosystem that collectively mobilises engineers, researchers, students, healthcare professionals, industries, regulatory authorities, government decisionmakers, and patients to rapidly generate ideas and technological solutions. As a living lab, iTMT offers a place for all stakeholders to discuss their concerns, to support research, validate, and accelerate innovation (Mérindol and Versailles, 2017).

As a type of open lab, the living lab approach favoured by iTMT is very much like the definition of Dubé et al. (2014: 12) an "Open innovation method aimed at the development of new products and services. The approach promotes a process of co-creation with end users in real conditions and is based on an ecosystem of public-private-citizen/user partnership". The agility of iTMT's living lab makes it possible to associate early in the development phases the various stakeholders, including users who will be affected by the technology to mobilise their knowledge in real time (Berthou and Picard, 2017; Dubé et al., 2014; Pallot et al., 2010). The involvement of different talents playing key roles such as knowledge brokers, boundary spanners, energisers, or various intermediaries capable of coupling different transdisciplinary

DOI: 10.4324/9781003125587-13

knowledge sets, proved to be a real asset for iTMT (Fleming and Waguespack, 2007; Meyer, 2010).

iTMT's main mission is to co-develop and validate next-generation medical technologies intended for diagnosis, prognosis, interventions, and rehabilitation applied to three major disease areas: cancers, cardiovascular diseases, and musculoskeletal diseases. In March 2020, as new priorities emerged, iTMT responded to an urgent call from the provincial government and medical institutions to launch an *Innovation Project Grant* to meet the needs generated by the pandemic (Covid-19).

The institute was one of the first initiatives in Quebec in the domain of healthcare to envision the entire innovation value chain from the research project to the premarket field by focusing the innovation process on the users,[1] bringing together all stakeholders to co-develop the technology. Furthermore, iTMT also addresses some of the industry's main preoccupations: research and industry alignment by implementing quality management systems for the regulatory standards for medical device development (ISO 13485:2016)[2] and addressing the potential challenges of intellectual property (IP) in an open innovation context, and identifying potential markets at an early stage.

In this chapter, we aim at analysing the emergence and the development of iTMT. From such a perspective, we highlight the role of a community of passionate researchers who succeeded in establishing a shared vision at the origin of the living lab. We then focus on the specific dynamic of innovation that emerged from the living lab mode of organisation adopted by iTMT and show that this dynamic is characterised by sequences of community formation and the orchestration of diverse "commons" to foster and validate innovation.

This contribution is based on a case studied from two complementary angles (Eisenhardt, 1989; Yin, 2009), from 2019 to 2021, where data were collected mainly through 100 semi-structured interviews with strategic iTMT employees, stakeholders, partners of iTMT, and with public representatives and members of medical technologies and innovation communities in Canada and in many countries. The analysis was completed with an extensive scientific literature review.

9.1 Theoretical background

Our view is that the understanding of the creation, development, and evolution of iTMT is inherently related to the articulation of different *commons* in order to establish interdisciplinary boundary-crossing collaboration. In line with Ostrom's seminal approach, we consider *commons* as "collective action governance mechanism over a common pool resource shared by the members of a community who jointly manage the use and access to this resource as well as its preservation or development" (Zimmermann, 2020: 106). Based on our observations, the dynamics of iTMT as an active living lab in healthcare focuses on the formation, development, and maintenance of the following sequence of *commons*: (a) *social commons*, (b) *symbolic commons*, (c) *innovation commons*.

9.1.1 Social commons

The *social commons* are a collective response to imperatives related to the essential needs of social groups such as health, employment, culture, education, etc. (Defalvard, 2017). As a category of social commons, the *social commons* (Helfrich et al., 2009; Simon et al., 2021) are a common pool of resources based on the formation and maintenance of different forms of "know-who" (know-who shares the same interest, know-who has the competencies, know-who can help, know-who knows who, etc.).

9.1.2 Symbolic commons

The *symbolic commons* aim at framing the purpose, legitimacy, and focus of a collective endeavour. Such *commons* give the directions for the investigations and experimentation between stakeholders. This symbolic commonality leads the co-creation of a *manifesto* in the form of a declaration of purpose, which fosters common vision, commitment, and desire to cooperate.

9.1.3 Innovation commons

Building on the notion of *knowledge commons* highlighted by Hess and Ostrom (2007), Potts (2018, 2019,) and Allen and Potts (2016) introduces the notion of the *innovation commons*, as a rule-governed space for solving an important problematic or discovering opportunities inherent in sharing tangible and intangible resource inputs into innovation. The *innovation commons* exist at the very beginning of a new technology, at the point when a group of actors (scientific and clinician, entrepreneur) come together to explore for the development of the technology to answer an unmet or important need. Allen and Potts (2016) point to the dynamic bases of distributed information and new knowledge assets that are accessible and exploitable by all the stakeholders, including entrepreneurs, of an innovation process to actualise an innovation. Their vision of the *innovation commons* also insists on the modes of collective action activated by the stakeholders not only to reap the benefits of the exploitation of the common pool resources of innovation, but also to nurture its development and ensure its sustainability.

Potts (2019) discusses implications for policymakers who seek to jump-start innovations. He highlights that many attempts will focus too heavily on later phases of the innovation trajectory, neglecting the crucial first stage. He suggests that a better approach to innovation would be to facilitate collective learning upstream. In his perspective, the knowledge pool is a combination of things and ideas. Within *innovation commons*, new technologies, adaptations (by the user) of existing technologies, and ideas on new ways of using these technologies are shared. He challenges the heroic Schumpeterian entrepreneur and replaces it with the idea of vibrant communities and clubs of experimenters (for example: Homebrew computer club). For Potts (2019:

46), "Innovation is something we do together. It takes an economy to develop a technology".

For each of the *commons*, following Ostrom's approach, we paid great attention to the community at the origin of the *commons* which manages the use and access to the common pool resource. As Helfrich et al. (2009: 9) underline,

> Managing common pool resources in a practical sense requires a community that becomes aware of its relationship to the resources in a social context and names the resources as its own– a community that claims them, a community that presses for and helps, enforce rules to respect this co-ownership.

Our analysis, detailed in the following sections, highlights that the sequence of *commons* is the result of a specific dynamic of interactions of different communities which is orchestrated by an informal group of passionate scholars and medical professionals.

9.2 The dynamics of the creation of ITMT

9.2.1 The origins of iTMT

The iTMT initiative was launched by a group of leaders belonging to different institutions in Montréal. Prominent actors led the way, most from joint Institute of Biomedical Engineering (Polytechnique Montreal, UdeM and its affiliated university hospitals), but also from the Segal Cancer Centre of the JGH.

All members of this community had an excellent reputation among their peers and had access to a wide network of academic and industrial talents. They were convinced that to develop the next generation of medical technologies for complex diseases, a new interdisciplinary approach was needed in the healthcare system. The leading researcher-professor of Polytechnique Montréal and co-creator of iTMT was introduced to the living lab approach during his doctoral studies, 30 years ago. At the time, the notion of a living lab was not a defined concept, but the *principles* were used to develop and test technologies in a real-world context, to foster collaborations between users and developers, and to answer the medical needs. Since then, Polytechnique Montréal engineer-researchers are therefore incentivised to collaborate with physicians when developing medical technologies.

The vision elaborated by the pioneering researchers of iTMT and the capacity to attract new members to join the community was facilitated by the credibility of the founding members in their respective research domains and through their collaborations with the clinical world. This acquired legitimacy was further reinforced by their capacity to convince and embark more than 30 partners who are key players ranging from health institutions, academic communities, public funding agencies, or business and industry sectors.

Priorities emerged from this collective initiative. The first was the use of the living lab approach to bring together all the stakeholders who needed to support the entire innovation value chain from the research to the implementation of innovation. The second was to mobilise researchers to focus on strategic areas, to develop vocations among graduate students, and to attract the best talents locally and internationally.

9.2.2 Elaborating a common vision: the role of the manifesto

In 2015, the founders of iTMT and their partners applied for their first major grant from the Canada First Research Excellence Fund (CFREF). The first request was rejected. However, this failure allowed iTMT stakeholders to use the following year to strengthen their partnerships, to add new collaborators, to refine the living lab approach, and to support the unfolding of a sequence of *commons* development, as explained in the next section. Because iTMT was a pioneer in the establishment of a new organisational form in the healthcare milieu, the second proposal took the shape of a quasi-manifesto, asserting a genuine orientation with strong operating principles and shared values.

The iTMT co-founders had a desire to change the linear innovation model to adopt a co-creation approach in real-life context. They benefited from this period to clarify the modes of governance to validate their living lab founding principles:

- Meeting a need identified by users.
- Finding a solution that did not exist elsewhere.
- Designing technologies considering regulatory constraints.
- Integrating training into practices.
- Co-creating with stakeholders.
- Promoting interdisciplinary and transdisciplinary collaboration.

To support this vision, they created a scientific steering committee, a governance committee on which sit expert representatives of the community. Following this, they formalised a collective project approval process, confirming the founding and distinctive strategic activities and principles. The following year the co-founders resubmitted and received the grant. The Institute was founded in September 2016 and is described by the CEO as a "catalyst for the development, intermediation, and mobilisation of innovation".

Since then, iTMT's mission is defined as to support the development of next-generation medical technologies for cardiovascular, neuro-musculoskeletal, and cancer diseases with the aim of facilitating their implementation in the health system, training the next generation of medical technology professionals, and making innovation in life sciences and engineering an asset for society.

iTMT is focused on five main lines of activities. They offer scholarships and training pathways for students. They support R&D through the recruitment of professors and the creation of research chairs. They help with the financing of technological platforms supported by highly qualified professionals (HQP), and they accompany numerous projects related to scientific research in innovative medical technologies. More recently added, they are supporting entrepreneurship partly through the *Innovators in Residence* programme, and they engage in international development and outreach activities.

9.2.3 The institutionalisation of iTMT

iTMT is currently within the CHUSJ in Montréal (Canada). The choice to locate the institute within a university health centre (CHU) was a strategic decision that aligned with the shared vision of the founding members to create a living lab. Moreover, the leader and co-founders of iTMT had developed privileged relationships with the medical teams and the hospital management over the years. This clinical exposure and the strong relationships established gave iTMT managers a unique position: they had both an innate understanding of the challenges and practices of the medical institutional environment, and a privileged access to medical teams and facilities. In this open innovation perspective, specialists from different disciplines, students, patients, caregivers, healthcare system actors were engaged in a collaborative process to foster the co-development and implementation of medical technologies that address important unmet clinical needs. The goal was the acceleration of the integration of these transformative technologies into the healthcare system.

During the creation of iTMT, the CEO of the CHUSJ had a strong focus on innovation as a key value of the institution. The CHUSJ had also adapted its organisational model, its philosophy, and its management practices according to the principles of agile organisations responding to the constant fluctuation of the environment (Denning, 2016) with the introduction of communities of practice, elements of lean management, and a progressive refocus on the patient's journey and experience. The objective of this model for the CHUSJ was to increase the organisation's ability to react to the evolving environment and to mobilise its teams to achieve the hospital's mission: care, teaching, and research (Brunet and Malas, 2019).

Beyond the CHUSJ's boundaries, iTMT has progressively anchored itself in the vibrant Montréal healthcare ecosystem, which includes other university hospitals such as the CHUM and the JGH, and universities (UdeM with an important faculty of medicine, and Polytechnique Montréal, a university specialized in engineering). The more iTMT seeks to bring together major stakeholders interested in being part of processes that accelerate technology, the more it energises the Montréal and provincial healthcare ecosystem.

In this regard, following Clayton, Feldman, and Lowe's (2018) definition, iTMT can be seen as an *innovation intermediary* in this ecosystem by facilitating the process of developing and commercialising scientific innovations. Indeed, since academic entrepreneurs often possess limited business skills and experience, the role of intermediaries is to assist them by counterbalancing the information asymmetry that can hinder their innovation process. These intermediary organisations thus offer support for innovation through a variety of services ranging from financing to mediation, advice on business models, evaluation of potential market, intellectual property rights and protection, regulation, and international certification. When the need is expressed by users, iTMT delegates, and coordinates with stakeholders to gather expertise and resources needed to validate, develop, and implement innovative solutions. Its support for the stakeholders involved has made it possible to mitigate risks and, sometimes, to accelerate the process of developing innovation into pilot projects before going to the market.

9.2.4 Role and contribution of iTMT living lab: a catalyst of innovation

In all its spheres of activity, iTMT adopts the living lab approach to promote transdisciplinary, intersectoral research, and open innovation. This approach allows all stakeholders (engineers, healthcare professionals, students, patients, industrial partners, hospital administrators, etc.) to participate from the early stages of research to the implementation. At the beginning of the project, stakeholders actively co-develop solutions, challenging their ideas to optimise future development, validation, commercialisation, and processes. Stakeholders will simultaneously consider the aspects and performance of the medical technology or intervention process and its potential adoption by users. Thanks to meetings with regulatory experts, market analysts, IP professionals, medical device evaluators, user groups, etc., research teams emerge with a complete roadmap for the development, implementation, and commercialisation of their technology. For partner firms, iTMT quality project management (ISO 13485: 2016) also mitigates the risks and consolidates the increased value of the IP.

iTMT is an open, transdisciplinary, and intersectoral innovation ecosystem, driven by user needs (patients and their families, community, healthcare professionals, healthcare system managers, technical professionals) and by the dynamics of needs (push/pull/pool/partnership), conducive to creativity and innovation, as well as their validation and implementation in the healthcare community.

The living lab approach is not only a way of exchanging knowledge and ideas between partners, but also a way of being and socially interacting. Researchers and students who experienced the living lab approach through iTMT activities, decided to adopt and implement this approach, to their home laboratory, firm, or medical institution. They added, “This implementation of living lab mindset was not always successful, but above all, the social interaction between the staff of their home organisation was profoundly changed, in a resolutely more open framework of exchanges and positive human relations”.

Beyond this conceptual outlook, iTMT managed to foster the enrolment and onboarding of multiple communities under a common, overarching vision, and through constant training, knowledge development, and knowledge sharing activities. The constant support provided by iTMT to researchers, graduate students, or highly qualified personnel contributes to the training of expert skills essential to the production of knowledge, but also to the maintenance of scientific and technical platforms that are now essential to advanced research practices. From this point of view, the institute and its leaders have not only been able to create a truly dynamic ecosystem around their activities by connecting different researchers, experts, entrepreneurs, and industrialists, but also by triggering and facilitating interactions between different local and international knowledge communities.

9.3 The orchestration of a sequence of commons as the cornerstone of the development of iTMT

Based on the above empirical observations, our interpretation of the dynamics of emergence and development of iTMT as an active living lab in healthcare, focuses on the formation of a sequence of *commons* orchestrated by an informal group of passionate and visionary scholars and medical professionals to establish interdisciplinary boundary-crossing collaboration. We propose them as the "commoners" (according to Ostrom, the members of the community who care about the *commons*). Aiming at developing a vibrant living lab within a hospital, the main objectives and cognitive mechanisms of the commoners changed through time as the project of the living lab matured. They got involved in different types of communities to be able to gather the ideas, the skills, and the diverse types of knowledge required for such a challenge. Consequently, the commoners had to focus on the use, access, preservation, and development of the following sequence of *commons*.

9.3.1 Social commons and the professional community

At the origin, the collective initiative started from an informal group of scholars and medical professionals belonging to different institutions in Montréal sharing the same professional interest: they were convinced that to develop the next generation of medical technologies for complex diseases, a new interdisciplinary approach and co-design were needed. Such a community brought together many of the characteristics of a "professional community" as defined by Amin and Roberts (2008: 257) in their analysis of the different varieties of knowing in action: specialised expert knowledge acquired through prolonged periods of education research and training, institutional trust based on professional standards of conduct, sharing of the same experiences and values, interest for radical innovation stimulated by contact with other communities, etc.

As for any professional community, this one was structured along different levels of engagement from its members: a core group of very active members, some regular members, and many browsers (Wenger, 1998). To a large extent, the core group of the community corresponds to the *commoners* previously described. This core group wanted to increase the potential of knowledge sharing and collective learning of the wider community by promoting a spirit of solidarity between the stakeholders.

This translated as the development of a *social commons* (Helfrich et al., 2009; Simon et al., 2021) mainly focused on the active opening and intersecting of personal networks. More concretely, the main objective of the professional community was that more experienced stakeholders would often facilitate the work of younger scientists by connecting them with potentially helpful contacts from their personal networks. In this community, the focus was on co-building strong *social commons* based on *know-who* (Lundvall and Johnson, 1994).

This priority to sustain the *social commons* was clearly highlighted by the core members of the community. As we will see in the following sections, the emergence and development of iTMT will imply the orchestration of new types of communities and new types of associated *commons*, but the formation and the nurturing of the *social commons* could be considered as the core element of the foundation of the iTMT living lab. On a day-to-day basis, the commoners continue to take great care of the preservation and continuous enrichment of the *social commons*, under their common, overarching challenge.

9.3.2 Symbolic commons and the epistemic community

Once the professional community reached an active level of development, the commoners started considering switching from "thinking" to "creating". They progressively co-constructed a shared vision aimed at implementing a living lab equipped with the values of knowledge sharing and collective learning, and located within a health centre, preferably a hospital. This symbolic commonality leads to the co-creation of a *manifesto* in the form of a declaration of purpose, expressing the breaking of the rules of the traditional ways to develop R&D in the healthcare domain. Such a *manifesto*, which fosters common vision, commitment, and desire to cooperate, was the basis of the application for the major grant from the CFREF.

In many respects, this new community initiated by the commoners could be qualified as an *epistemic community*. As defined by Cowan et al. (2000: 234), epistemic communities are small groups of "knowledge-creating agents who are engaged on a mutually recognized subset of questions, and who (at the very least) accept some commonly understood procedural authority as essential to the success of their collective activities". As a key element of the procedural authority, the *manifesto* helps attract towards the *epistemic community*, not only members of the initial professional community, but also influential members from the top management of hospitals and other medical organisations, as well as industrialists and policymakers, who share the values and visions of this community.

The *epistemic community* contributed to the development and nurture of the *symbolic commons* based on the intent to frame the purpose, to give the directions for the investigations and experimentation of the different members of the community. The *symbolic commons* that progressively emerged from the interactions between members of the community helped in diminishing the centrifugal forces and to constantly refocus the discussions of the participants under the overarching challenge of this innovative co-development project in the healthcare sector. The symbolic commons also played a key role as a visibility factor for all health institutions and for patients. This capacity to create symbolic commonalities has been made possible by the legitimacy acquired from the outset thanks to the reputation of the team of researchers who initiated iTMT, but which has also been reinforced by the capacity of the community created around the institute, to convince and onboard important players in the industrial and business worlds, government services, or health institutions. An equally important spin-off of this legitimacy is to give confidence to potential donors and investors, both public and private, in projects that often seem too risky to finance.

It is worth noticing that the common vision at the origin of the *symbolic commons* is not fixed and given once and for all. It is a dynamic process shaped by the interactions and debates between participants and by the changes in the economic and social environment. As an example, the crisis related to the pandemic contributed to the introduction of some changes in the common vision of the community, with an increasing emphasis given to the security of carers and patients, or on the growing need to develop rapid and sometimes frugal innovative solutions.

9.3.3 The development of the innovation commons orchestrated by iTMT

The institutionalisation of iTMT in 2017 could have led to a clear transition from a community-based initiative to a traditional formal way of managing (with a hierarchical structure, obligations to achieve results, contractual rules, and procedures, etc.). However, the imperative to build a living lab as a "catalyst for the development, intermediation, and mobilisation of innovation" (according to the CEO of iTMT) rapidly convinced the iTMT team that a hybrid form of organisation was fundamentally needed. Consequently, it was intended to combine the formal structures of iTMT with a community-based mode of functioning to establish interdisciplinary boundary-crossing collaboration to find innovative solutions in healthcare.

Drawing on the initial impulse from the commoners, and on the already existing *social commons* and *symbolic commons*, iTMT started developing a series of *commons*, that we call *innovation commons* with reference to Allen and Potts (2016) to foster innovation.

These *innovation commons* can be defined as dynamic bases of knowledge assets that are accessible and exploitable by all the stakeholders revolving around the living lab and maintained and enriched by collective debates and feedback from their exploration and exploitation. These innovation commons exist from the beginning for the use and development of the adapted technology. Allen and Potts (2016) call them *innovation commons* because the common pool of resources is not the technology *per se*, but information and knowledge about the technology that subsequently facilitates its development and its transformation into an actual innovative solution.

The notion of *innovation commons* highlights the importance of the collaborative management of common pool resources (especially knowledge resources) based on a set of rules that are self-produced and accepted or adopted by all stakeholders to foster the dynamics of innovations. The institutional arrangements elaborated by iTMT encourage the sharing and mutualisation of human, scientific, technological, financial, and social resources. iTMT constantly strives not only to support the production of knowledge (experiments, publications, communications, etc.), but also, and above all, to develop dynamic and accessible repertoires of various types, shared between stakeholders, where the beneficiaries agree to give each other mutual access to the knowledge they hold. In practice, stakeholders are constantly invited to share the learning specific to one project with the other stakeholders of the different communities involved in iTMT.

Such *commons* were not limited to the *traditional actors* in the health innovation process but are open to any discipline or sector potentially concerned. Innovation intermediaries must, for example, develop specific common institutional arrangements to make it possible, among other things, to determine the roles and responsibilities of each party, the rules of access and appropriation of the resources (cognitive, social, material, financial), that are jointly developed or ways of validating and legitimising new ideas with other external stakeholders.

The results of interviews conducted with iTMT stakeholders, confirm that the institute is in line with the most advanced theories and practices in open, participatory, and collaborative innovation, through its ability to alleviate uncertainty around a nascent technology by pooling distributed information about uses, problems, and opportunities. The adoption of international practices and standards (ISO) also play an important role in risk reduction and value creation. These conditions allowed iTMT to create multiple *innovation commons* for the benefit of the participants in iTMT's activities. More precisely, the analysis of iTMT management shows that in a few years of operation, iTMT has contributed to the creation and development of the following *innovation commons* (Potts, 2019).

9.3.3.1 Technology/product/service commons

This category of *commons* encompasses the knowledge useful to the entrepreneur/innovator in the development of a technical and functional innovative solution. According to Allen and Potts (2016), "This is the first and most obvious (resource) being technical knowledge (and associated physical resources) that describe the new idea or technology", leading to *Technological-Scientific Discovery*. We identified three subcategories of knowledge here:

1. *Technical/technological commons*: the knowledge needed to design and prototype a functional solution. For example, iTMT's commitment to the development of scientific and technical platforms ensures the establishment, exploration, and operation of real strategic technological

matrices for the advancement of research. The provision by iTMT of HQP within technological platforms makes it possible to support the implementation of interdisciplinary and intersectoral projects for the development, validation, and implementation of medical technologies with high potential impact for the healthcare system and the patients.

2. *Skills commons*: even if this would quite often be underestimated, this is the shared *know-how*, the procedural applied knowledge needed to support key activities such as platforms in the iTMT living lab. These elements of knowledge may not be directly linked to the main knowledge base and the fields of expertise of iTMT, but they appear as key elements to leverage the collective efforts towards innovation. For instance, iTMT, through experiences and linkages, developed a specific expertise in project management, research grants, intellectual property, market potential, business model, intersectoral innovation opportunities, or technology transfer. This knowledge is extremely contextual and specific, difficult to acquire and to imitate. It is certainly a key element for the translation of a technology concept into a well-funded and developed the project.
3. *Common technological platforms*: the recent evolution of innovation models saw the emergence of innovation practices based on the exploitation of technological platforms, around which sophisticated ecosystems can develop (Gawer and Cusumano, 2014). These platforms can be physical, for instance in medical devices or materials (microfluidics, imaging, etc.), or digital, as in the internet, medical data or more recently, blockchain (de Reuver et al., 2018).

In line with Potts, we believe that these elements of *knowledge commons*, if necessary to innovation, are not sufficient to support its development. They should be complemented by the following *commons*.

9.3.3.2 Entrepreneurial commons

Again, as emphasised by Allen and Potts (2016: 2), another category of knowledge is needed:

> A second type of resource in the innovation commons that could be largely invisible to the uninitiated: The distributed bits of information which are taken together, help define the entrepreneurial opportunity associated with that new technical idea that leads to entrepreneurial-market discovery.

This type of knowledge relates to strategic insights about the environment, marketing intelligence, market structure, competition, distribution channels, or regulatory issues, for instance. It appeared that iTMT's deep understanding of the functioning of public hospital tenders proved challenging, especially for young scientific entrepreneurs and industrialists. Here, the experience of medical teams, hospital managers, and experienced researchers helped to orient innovation projects and further define their business model. A lawyer with an extended expertise in the valorisation of innovation projects in medtech and healthcare is also available to support the orientations of entrepreneurial endeavours and disambiguate issues of IP and commercialisation. The ongoing interactions of iTMT members with entrepreneurs and the constant connections with local incubators and accelerators allowed them to introduce tools and methods to researchers and scientists from entrepreneurial practice and to validate and consolidate the commercial and economic viability of projects.

9.3.4 The key role of the commoners

The interpretation in terms of *innovation commons*, which has just been developed regarding the functioning of iTMT as an intermediary innovation organisation, underlines the distinction between the different forms of *innovation commons* that are at the basis of a dynamic living lab approach. It also highlights the importance of the generative capacity that results from the coupling and energising of the different forms of *innovation commons*. iTMT's organisational resilience and agility implemented in response to the pandemic crisis, in an accelerated mode, demonstrated the potential for value generation, and the reduction of the risks.

However, besides these results, the analysis has also emphasised the key role played by the informal group of passionate scholars and medical professionals called the *commoners* in this chapter. The *commoners* got involved as a core group in different types of communities to be able to gather the ideas, the skills, and the diverse types of knowledge required for developing a living lab. As each community focused on a given *commons*, the *commoners* took care of the use, access, preservation, and development of the sequence of diverse *commons*.

As defined by Maguire et al. (2004: 657), iTMT commoners can be seen as institutional entrepreneurs since they "have an interest in particular institutional arrangements and [can] leverage resources to create new institutions or to transform existing ones".

The *commoners* must constantly encourage the sharing of knowledge between the different stakeholders, either through codification and combination, or through intensive socialisation. They also ensure that knowledge exploitation by a specific stakeholder would always be conditioned by some of the following returns – economic subsidies, shared learnings, new contacts, or even in terms of reputation – for the different communities involved, to support the continuous enrichment and regeneration of the common pooled resources. This proves possible because most of the stakeholders believe in the *symbolic commons* and engage to address the overarching challenge.

From an ecosystem perspective, this also raises significant challenges to maintain the sustainability of the *model*. The limits and the traps are multiple, in particular the risk of the "cognitive exhaustion" of the *commoners* as in many communities ("There is no more grain to grind", "We have done all we can do", etc.). Another challenge is to keep the agility of iTMT and to prevent too much institutionalisation or bureaucratisation without flexibility. An implementation of rules from the top down could kill the different *commons*.

Another risk is that the different communities that support iTMT (professional community, epistemic community, and the different communities of practice attached to the *innovation commons*) become too large and unmanageable. Part of this can be mitigated by a constant effort to support knowledge sharing in and between communities and projects and reinforced by regular events that bring the different communities together.

Therefore, though successful and performing, the living lab model of iTMT is fragile and requires the constant attention from the *commoners*.

9.4 The *Innovators in Residence* programme: an example of *entrepreneurial commons* generated by iTMT

One of the examples of *entrepreneurial commons* generated by iTMT that we explored in depth through action research is the Fonds de Recherche Québec *Innovators in Residence* (IiR) programme hosted and led by iTMT. The objective of this programme is to accelerate the pre-commercialisation process of a technological innovation in healthcare developed by a local start-up by providing access to its relevant clinical environment to test its technical and institutional implementation conditions. Coming from an increasing need, the goal of IiR was to improve collaboration between innovators and the healthcare sector and to support the entrepreneurial process until the technology is validated with actual users in the actual context.

Pairing the entrepreneur with a junior researcher, the innovators (as an example, the first pairing was an entrepreneur in genomics and junior researcher in management) were closely supervised via an operational on-site committee (with local managers and healthcare professionals from the hospital) and monitoring (managers from the hospital, with external innovation experts and members from funding bodies).

The living lab practice of iTMT allows for a constant mobilisation of diverse stakeholders in support of the project, accelerating on-site experimentation and validation. With the financial and project management support from iTMT, innovation processes are accelerated, and knowledge generated for all stakeholders. The monitoring committee made sure to connect the entrepreneur with the relevant services to move forward on his/her journey: ethics office for clinical research protocol, procurement department, legal department, connection with experts in technology assessment.

9.4.1 Exploitation of commons

iTMT graciously offered to install the first cohort of *Innovators in Residence* in its facilities within the hospital, thanks to the support of CHUSJ's top management, who were convinced of the added value of the IiR programme. The key hypothesis was that the development and validation of the innovation would have more chances to succeed on the site, with the possibilities of direct access to the whole array of potential users and stakeholders. It allowed the start-up's employees to be part of the research committee and see how clinicians and researchers in bioinformatics interacted with genomic information. It was also possible for the innovators to consult the biomedical technology data analysis software to help them carry out their market research. However, iTMT and the entrepreneur realised quickly that tacit rules, governance principles, and commonalities needed to be defined as a basic starting point with all stakeholders. iTMT's premises is not only an open physical space, but also a space for knowledge sharing. The living lab is a real-life experimentation environment in which new products and services are developed through the collaborative efforts of users and developers, using various resources to carry out the project (Hakkarainen and Hyysalo, 2016).

9.4.2 Generation of commons

One of the objectives of the IiR programme was to document this exclusive support process for the validation of a technology from a healthcare start-up. In an interdisciplinary mode, it was one of the main tasks of the junior management researcher to trace the difficulties and solutions found during the validation process: *what were the crucial resources? were some of them missing? how to ensure that adequate experimental conditions will be reproduced in the future? what institutional constraints are to be expected and how to deal with them?*

The success of the first edition of this programme IiR made it a testbed and a proof of concept for iTMT institutional project, to accelerate innovation within hospitals and particularly to support technological entrepreneurship in this environment. For iTMT, this experimentation became a demonstrator of their operating principles as a living lab. Every time visitors came by in iTMT's open space, it was an opportunity for the iTMT management team to tell the story of the IiR programme and the evolution of the start-up growing *in vivo*. Internally, it also encourages the community of life science researchers and students to consider entrepreneurship as an option and ask iTMT for its expertise to help them in their endeavours. From now on, one of their main areas, an activity called "entrepreneurship", is entirely focused on supporting this burgeoning community. Since this first successful pilot, two other editions have materialized in other partner hospitals and with students from other universities with even greater repercussions.

This entrepreneurship support programme illustrates, on the one hand, the articulation of *symbolic commons* with the *innovation commons*, as described by Potts (2018). Indeed, this articulation of commons in the form of hybrid arrangements allows the structuring of *entrepreneurial commons*. However, the question of the sustainability of this programme as *entrepreneurial commons* may arise. The relationships between the programme's "godfathers" are solid, their objectives are aligned, and their initiatives are supported by higher authorities. On the other hand, since the goal is to contribute to the development and commercialisation of innovation, should these *entrepreneurial commons* be structured by strict intellectual property mechanisms? This raises the following question: *how can we ensure that entrepreneurs benefiting from this type of support contribute to the maintenance of the entrepreneurial commons*? The literature on entrepreneurship is full of examples showing that the entrepreneurial process pushes entrepreneurs to be more opportunistic (Coase, 1988) – as seizing opportunities and exploiting conditions (Sarasvathy, 2009) – than grateful by following the logic of counter-giving (Mauss, 1950). Two arguments could go against this. The first one could be ethical, hoping that the entrepreneur sees the *symbolic commons* as meaningful enough to justify sharing a part of the revenues of the entrepreneurial venture as a benevolent, philanthropic gesture. The second one, more pragmatic, would argue that the value of belonging to the living lab community is higher than the short-term value of exploiting its assets. If the entrepreneur only exploited the knowledge from the lab, without giving back (in many possible forms: new knowledge, new fundings, new connections ...), they would be at risk of being ostracised by the community, hence losing all the former benefits of participation. In this regard, the challenge of the lab is to demonstrate that it could provide value to the entrepreneur in the long run to continue improving his/her project, even after launch, or for future endeavours.

The development of this *entrepreneurial commons* orchestrated by the IiR programme within iTMT makes it possible to establish a truly open, dynamic, and shared *space* that promotes the establishment and strengthening of a genuine value chain of collaborative and integrated innovation, articulating knowledge production, generation, experimentation and validation of new ideas, and concrete development of innovative projects. Numerous mechanisms are designed to foster the development of entrepreneurial activities based on iTMT's activities.

One of the mechanisms is that of shared reflections on issues related to property rights. One of Ostrom's major contributions is not to have restricted the question of property rights to the exclusive property aspects of a right to a particular actor (for example, patents granted to an innovator), but to consider that in the construction phase of the *commons of innovation*, what is important is to specify the nuanced modes of use of property issues between the different actors (*can one access the resource? can one use the resource, if so under what conditions? can one of the actors exclude the others from using the resource? is the resource transferable? how can it be maintained, nurtured, enriched? etc.*). The pooling of sufficient information and collective resources to reduce, among other things, uncertainty and enrich the full potential of innovative ideas and concepts. Once these pooled resources have been co-constructed, the validation of entrepreneurial solutions, the deployment of commercial solutions and the implementation of applications can be achieved much more efficiently than if only the technical, scientific, or commercial aspect of the innovation problem had been prioritised.

In this perspective, one of the advantages highlighted in the interviews and resulting from iTMT's living lab activities is to enable projects to better cross or mitigate the "*Death Valley*" of innovation (Ford et al., 2007). Since the path from a discovery resulting from basic research to a commercial product or process is long and exposes itself to the difficulty of implementing, accelerating, or commercialising an innovation project, a particularly difficult period is when a technology is not yet mature enough for industry but is already too applied to be academic. Innovators and investors thus regularly assert that there is a transition phase between basic research and the commercialisation of a new product, known as the *Death Valley*. This term is used as a metaphor to describe the relative lack of resources and expertise in this area of development. The metaphor suggests that there are relatively more resources on one side of the valley in the form of research literature and grants, and on the other side in the form of commercialisation expertise and resources and equity funding, but that there is a severe lack of resources in between, when prototyping and real-life, on-site experimentations are becoming key to ensuring the translation of a proof of concept into an actual functional innovation, equipped with a relevant business model.

While the CHUSJ was already very open to public-private partnerships, iTMT does not hesitate to reveal some constraining but implicit rules of a public hospital. For example, the rules applying to the hospital are sometimes blurred in the way they apply to the institute. These rules make it possible to stabilise the CHUSJ as a healthcare organisation as well as an R&D partner. However, promoting commercialisation processes and entrepreneurial programmes disrupts this desired balance by giving the perception that the commercial logic gains more weight. Indeed, health institutions that are often public must respond to several tensions or contradictory institutional logics (Reay and Hinings, 2009). Logics define what is appropriate and render certain actions unacceptable for organisations and individuals (DiMaggio and Powell, 1991) and represent culturally reinforced rules of action that have important roles in processes of organisational identity formation, sense-making, and legitimation (Suchman, 1995). In this way, Miller and French (2016) show how university hospitals (CHU) try to hybridise – to strategically coexist – the logics of healthcare and innovation.

In this phenomenon of hybridisation of logics, the authors point out that making an entrepreneurial shift would allow a CHU to: (1) better serve the hospital mission by giving priority to the needs of patients and to the organisation itself as a user of its own innovations and (2) achieve its goal of translational and effective health research, in as much as successful technology transfer and commercialisation is a transformative factor.

In a similar way (Wilden et al., 2018) presented the problem of the tension between focusing on the patient and involving multidisciplinary teams to provide the most personalised services possible and using fewer resources to contain healthcare costs. They characterised hybrid organisations as the organisational form, health entrepreneurs can create, to face institutional forces and absorb these tensions. As founded by actors with entrepreneurial profiles, iTMT seeks to absorb these conflicting logics by configuring its organisational form as a *hybrid*.

9.5 Conclusion

Through the analysis of the case of iTMT in Montréal, this study has highlighted the evolutionary dynamics of this unique living lab within a hospital in the healthcare ecosystem. The examination of the origin of iTMT confirmed the key and fundamental role played by a community of local actors from different universities and healthcare institutions in the greater Montréal region.

The study has also emphasised how the shared vision elaborated by the community progressively shaped and configured the different types of *commons* (*social, symbolic, knowledge, entrepreneurial/innovation commons*) that are at the core of iTMT living lab. Among the different *commons* developed by iTMT, a special emphasis has been placed on the *innovation* and *entrepreneurial commons*, through the analysis of the IiR programme to highlight the potential of hybridisation in giving priority to the needs of users, stakeholders, and to the organisation itself as a user of its own innovations, thereby facilitating successful technology transfer, commercialisation, and diffusion in the healthcare ecosystem.

As emphasised by Mérindol, Versailles and Le Chaffotec (2019), intermediation systems, such as iTMT, play a dual role, on networking (broker of networks) and on content (broker of contents). The logic of networking, intermediation systems allow the emergence of collective strategies between a variety of public and private actors. The various knowledge bases of these actors can be combined to foster and support a more efficient development of innovation. On content, they provide a variety of services that support the management of collective projects, or access to technological and real estate resources. Support for business creation is part of this second aspect.

To a large extent, iTMT can be considered as an actual "middleground" in the Montréal healthcare technology ecosystem (Simon, 2009, Cohendet et al., 2010).[3] Through fostering these sets of *commons*, the functioning of iTMT corresponds to all the elements constituting a *middleground*: *a "place"* which uses iTMT's user-friendly premises located in the heart of the CHUSJ; a series of "*events*" through the holding of regular colloquia, seminars, and events; a regular launch of new "*projects*" (on the development of solutions for coping with the pandemic, the *Innovative in Residence* programme, etc.); and the setting of "*spaces*" – the *symbolic commons* – as shared cognitive challenges and epistemic orientations, through the dissemination of white papers and newsletters are all mechanisms to facilitate interactions between stakeholders and shared learning (Sarazin et al., 2017). This pooling of *places* and *events* located in a real environment allows iTMT to include all participants in its activities to benefit from mutual access to various skills.

As a *middleground*, iTMT helps identifying, for example, promising R&D initiatives or already approved (licensed) equipment that can be reoriented and repurposed, then set up interdisciplinary teams to complete the expertise with scientific rigour until new protocols are approved and integrated into the healthcare system. The institute also allows beneficiaries of its support to have access to front-line clinicians in healthcare institutions to test innovative solutions *in situ*. This mutual access to skills between the various stakeholders also makes it possible to bring previously unidentified new talent around new projects, and to significantly increase the success of calls for tenders or grant applications, as well as the opportunities to develop start-ups.

Thus, iTMT is not just a living lab, but a living, inclusive, multisectoral hub, composed of players offering diverse expertise and coming from all sectors of the life sciences, medical technology, and healthcare sectors. Beyond its key functions of producing fundamental and applied knowledge, the institute is distinguished by its exceptional and structuring capacity to bring together a wide variety of actors around the implementation and deployment of the healthcare innovation value chain.

However, it is important to keep in mind that *innovation commons* are fragile (Potts, 2019). Certain *commons* are not immune from being captured by selfish private interests which might be tempted to appropriate the collective effort for themselves. The living lab approach taken by iTMT is not sheltered by the potential conflicts that can arise between actors, between communities or between groups. It is imperative that iTMT ensures that it continuously orchestrates proper functioning of the various commons and arbitrates the risks of conflict. Furthermore, there is a potential risk of conflict with institutions in healthcare innovation that still support the classical perspective, which refers to the linear model of innovation (Rogers, 2003). So far, iTMT has been able to be diplomatic enough and to convince prominent figures from these more classical institutions of the value of the living lab approach.

To conclude, Figure 9.1 encapsulates the different elements that have been highlighted in this chapter.

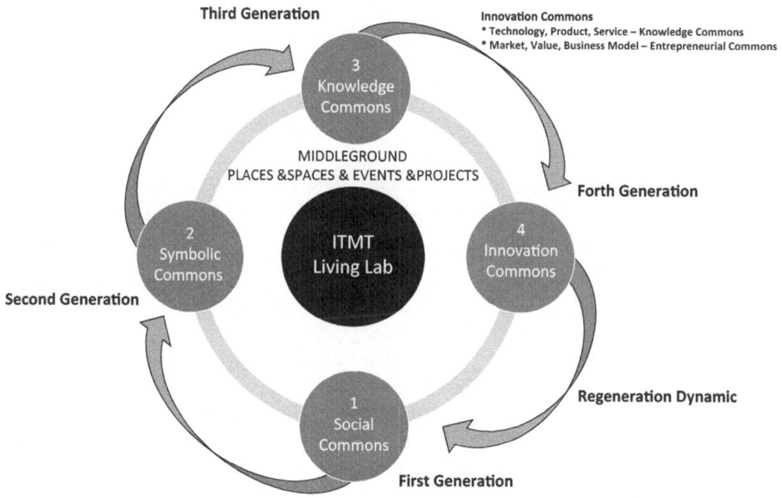

Figure 9.1 The dynamic of the activation of the commons.

Notes

1 For medical technologies, the users are primarily medical teams such as radiologist, pathologist, lab technicians, nurses, and sometimes directly the patients, etc. Nevertheless, the digitalisation of the healthcare system will imply more directly the patient with sensors and application.

2 In June 2020, the institute succeeded in obtaining the ISO 13485 accreditation ("*Medical Devices– Quality Management Systems–Requirements for Regulatory purposes*"). This quality standard aims to standardise research and development to offer laboratories supported by iTMT the possibility of accelerating the technology transfer as the guiding principles of this quality standard is also used in the industry.

3 Cohendet et al. (2010) define three different layers as the basic components of the creative processes in an innovative ecosystem: the *upperground*, the *middleground* and the *underground*. The *upperground* is the level of formal institutions (firms, government units, etc.). The *underground* is constituted by passionate individuals, creative talents, geeks, etc. In between, the intermediate layer, the *middleground*, enables new ideas to transit from an informal micro level to a formal macro level, through the accumulation, the combination, the enrichment, and the renewal of knowledge. The main elements of the *middleground* are *places*, *projects*, *spaces*, and *events*.

References

Allen, D., Potts, J. (2016) "How innovation commons contribute to discovering and developing new technologies." *International Journal of the Commons*, 10(2): 1035–1054. http://doi.org/10.18352/ijc.644.

Amin, A., Roberts, J. (2008) "Knowing in action: Beyond communities of practice." *Research Policy*, 37(2): 353–369.

Berthou, V., Picard, R. (2017) "Les living labs, ces leviers d'innovation en santé publique." *Annales Des Mines–Réalités Industrielles*, 2: 68–68. https://doi.org/10.3917/rindu1.172.0068.

Brunet, F., Malas, K. (2019) *L'innovation en santé : réfléchir, agir et valoriser*. Montréal, QC: Éditions CHU Sainte-Justine.

Clayton, P., Feldman, M., Lowe, N. (2018) "Behind the scenes: Intermediary organizations that facilitate science commercialization through entrepreneurship." *Academy of Management Perspectives*, 32(1): 104–124. https://doi.org/10.5465/amp.2016.0133.

Coase, R. H. (1988) "The nature of the firm: Origin, meaning, influence." *Journal of Law, Economics and Organization*, 4: 3–59.

Cohendet, P., Grandadam, D., Simon, L. (2010) "The anatomy of the creative city." *Industry and Innovation*, 17(1): 91–111. https://doi.org/10.1080/13662710903573869.

Cowan, R., David, P. A., Foray, D. (2000) "The explicit economics of knowledge codification and tacitness." *Industrial and Corporate Change*, 9(2): 211–253.

Defalvard, H. (2017) "From social commons to the commons society." *RECMA*, 345: 42–56. https://doi.org/10.7202/1040794ar.

Denning, S. (2016) "How to make the whole organization 'Agile'." *Strategy and Leadership*. https://doi.org/10.1108/sl-06-2016-0043.

de Reuver, M., Sørensen, C., Basole, R. C. (2018) "The digital platform: A research agenda." *Journal of Information Technology*, 33(2): 124–135.

DiMaggio, P. J., Powell, W. W. (Eds.). (1991) *The new institutional in organizational analysis*. Chicago, IL: The University of Chicago Press. http://catdir.loc.gov/catdir/toc/uchi051/91009999.html.

Dubé, P., Sarrailh, J., Billebaud, C., Grillet, C., Zingraff, V., Kostecki, I. (2014) *Le livre Blanc des Living Labs*. Montréal, QC: Umvelt Service Design.

Eisenhardt, K. M. (1989) "Building theories from case study research." *Academy of Management Review*, 14(4): 532–550. https://doi.org/10.2307/258557.

Fleming, L., Waguespack, D. M. (2007) "Brokerage, boundary spanning, and leadership in open innovation communities." *Organization Science*, 18(2): 165–180. https://doi.org/10.1287/orsc.1060.0242.

Ford, G. S., Koutsky, T., Spiwak, L. (2007) *A valley of death in the innovation sequence: An economic investigation*, Commerce Department, Technology Administration, Phoenix Center for Advanced Legal & Economic Public Policy Studies, revision 2018.

Gawer, A., Cusumano, M. A. (2014) "Industry platforms and ecosystem innovation." *Journal of Product Innovation Management*, 3(3): 417–433. https://doi.org/10.1111/jpim.12105.

Hakkarainen, L., Hyysalo, S. (2016) "The evolution of intermediary activities: Broadening the concept of facilitation in living labs." *Technology and Innovation Management Review*, 6(1): 45–58. http://doi.org/10.22215/timreview/960.

Helfrich, S., Jörg, H. (2009) "The commons: A new narrative for our times." In: Helfrich, S. (ed.), *Genes, bytes and emissions: To whom does the world belong*. Berlin: Herinrich Boell Foundation (Die Gruene Stiftung), p. 15.

Hess, C., Ostrom, E. (2007) *Understanding knowledge as a commons: From theory to practice*. Cambridge, MA: MIT Press.

Lundvall, B., Johnson, B. (1994) "The learning economy." *Journal of Industry Studies*, 1(2): 23–42. https://doi.org/10.1080/13662719400000002.

Maguire, S., Hardy, C., Lawrence, T. B. (2004) "Institutional entrepreneurship in emerging fields: HIV/AIDS treatment advocacy in Canada." *Academy of Management Journal*, 47: 657–679. https://doi.org/10.2189/asqu.2010.55.2.189.

Mauss, M. (1950) *Sociologie et anthropologie*. Paris: Presses Universitaires de France.

Meyer, M. (2010) "The rise of the knowledge broker." *Science Communication*, 32(1): 118–127. https://doi.org/10.1177/1075547009359797.

Mérindol, V., Versailles, D. W. (2017) "Développer des capacités hautement créatives dans les entreprises : le cas des laboratoires d'innovation ouverte." *Management International*, 22(1): 5872.

Mérindol, V., Versailles, D. W., Le Chaffotec, A. (2019) *Répondre aux défis de management de l'innovation en santé, le rôle des dispositifs d'intermédiation en France*. Rapport d'étude commandé par Genopole en partenariat avec BiFrance. http://www.newpic.fr/newpicopendoc/newpic-genopole-rapport-innov-sante-a4-2p-highdef.pdf.

Miller, F. A., French, M. (2016) "Organizing the entrepreneurial hospital: Hybridizing the logics of healthcare and innovation." *Research Policy*, 45(8): 1534–1544. https://doi.org/10.1016/j.respol.2016.01.009.

Pallot, M., Trousse, B., Senach, B., Scapin, D. (2010). Living Lab Research Landscape: From User Centred Design and User Experience towards User Cocreation. First European Summer School "Living Labs", Inria (ICT Usage Lab), Userlab, EsoceNet, Universcience, Aug 2010, Paris, France. ffinria-00612632f.

Potts, J. (2018). "Governing the innovation commons." *Journal of Institutional Economics*, 14(6), 1025–1047. https://doi.org/10.1017/S1744137417000479

Potts, J. (2019) *Innovation commons: The origin of economic growth*. Oxford: Oxford University Press.

Reay, T., Hinings, C. (2009) "Managing the rivalry of competing institutional logics." *Organization Studies*, 30(6): 629–652. https://doi.org/10.1177/0170840609104803.

Rogers, E. (2003) *Diffusion of innovations* (5th ed.). New York: Free Press.

Sarasvathy, S. D. (2009) *Effectuation: Elements of entrepreneurial expertise.* Cheltenham and Northampton, MA: Edward Elgar.

Sarazin, B., Cohendet, P., Simon, L. (2017) *Les communautés d'innovation : De la liberté créatrice à l'innovation organisée.* Caen: EMS Editions. https://doi.org/10.3917.

Simon, L. (2009) "Underground, upperground et middleground: Les collectifs créatifs et la capacité créative de la ville." *Management International*, 13: 37. https://doi.org/10.7202/037503ar.

Simon, L. et al. (2021) "*Reassessing Innovation Ecosystems through the Lens of Multiple Commons: MOVIN'ON LAB, A Case-Study in Sustainable Mobility.*" Working Paper Mosaic. HEC Montréal.

Suchman, M. (1995) "Managing legitimacy: Strategic and institutional approaches." *Academy of Management Review*, 20(3): 371–610. https://doi.org/10.5465/amr.1995.9508080331.

Wenger, E. (1998) *Communities of practice: Learning, meaning, and identity.* Cambridge: Cambridge University Press. ISBN 978-0-521-66363-2.

Wilden, R., Garbuio, M., Angeli, F., & Mascia, D. (2018). *Entrepreneurship in healthcare.* New York, NY and Abingdon, Oxon: Routledge. ISBN 9780367734176.

Yin, R. K. (2009). *Case study research: Design and methods* (4th ed.). Thousand Oaks, CA: Sage.

Zimmerman, J. B. (2020) *Les Communs : Des jardins partagés à Wikipédia.* Paris: Éditions libre & solidaire.

10 Open Labs in the transition from Triple to Quadruple Helix

Insights from smart cities and healthcare innovation ecosystems

Valérie Mérindol and David W. Versailles

The goal of this chapter is to identify how open labs contribute to the deployment of new modes of governance based on the Triple and Quadruple Helix models, with illustrations in smart cities and healthcare ecosystems of innovation. This contribution is directly linked to the emergence of new paradigms for the management of innovation (see Chapter 1 and Afterword). New dynamics for innovation, for instance linked to the digitalisation of the economy and the increasing use of digital technologies and artificial intelligence, require new collective strategies and practices between universities and actors of basic research, firms, policymakers, and representatives of civil society. In this context, Etzkowitz and Leydesdorff (2000) claim the necessity of deploying strategies elaborating on the Triple Helix (basic research, industry, and policymakers) while Carayannis and Campbell (2009) consider that innovation requires the installation of a Quadruple Helix approach where civil society plays an additional, and original, role.

This chapter considers two innovation ecosystems as illustrations to compare the dynamics based on the Triple versus Quadruple Helix models: healthcare and smart cities. Both sectors exemplify these frameworks for the governance of innovation at the territorial level. They investigate new technological trends such as digital technologies, big data, artificial intelligence, or additive technologies for the manufacture of new devices or the installation of new services. Even though they respectively apply to urban areas versus highly regulated and processed healthcare ecosystems, these two sectors make it necessary to transform behaviours, organisations, and ways of working. Petersen et al. (2016) have already shown that the Triple Helix approach is key for the management of healthcare-related innovation. The Triple Helix model also contributes to support the transformation of urban areas into "smart" cities based on digital technologies, real-time data management, machine learning, and other sorts of "artificial" intelligence. Collaboration between local policymakers, industry, and scientists, remains central in smart city programmes (Leydesdorff and Deakin, 2011).

Scholars and policymakers tend to progressively shift from a Triple Helix approach to a Quadruple Helix perspective. They do not only acknowledge the key role of end users and citizens in the adoption of new technologies but,

DOI: 10.4324/9781003125587-14

also, for the co-conception of innovations adapted to their needs. Smart cities and healthcare ecosystems represent environments where the Quadruple Helix approach seems to be relevant to addressing the governance of innovation and the management of innovation projects. Both the Triple and Quadruple Helix approaches require brokers who contribute to the creation of bridges between the various spheres (or helixes). This chapter investigates how open labs perform brokerage functions to install, foster, or expand the dynamics of Triple and Quadruple Helix models in healthcare and smart city innovation ecosystems.

This research is based on eight cases of French open labs (see Table 10.1): five cases focus on the management of smart city projects; three open labs are

Table 10.1 List of cases: thematic orientations, locations, and missions

Open labs	*Creation*	*Location*	*Thematic and mission*
TUBA	2009	Lyon	Experimentation of new urban services based on digital technologies and open data
Plages Digitales	2011	Strasbourg	Coworking space dedicated to entrepreneurs promoting digital projects
La Fabrique	2014	Strasbourg	Makerspace managed by the entrepreneurs of Alsace Digitale present in Alsace Digitale
ICM Incubator cLLAPS	2007	Paris	Incubator and living lab managed by the research centre on brain and spinal cord diseases. Active on all healthcare domains linked to neurosciences.
thecamp	2017	Aix-en-Provence	Contribute to resolutions of problems with strong societal impact
Make ICI	2014	Plusieurs villes en France Montreuil	Network of makerspaces and coworking spaces dedicated to rejuvenation of industrial wastelands and to craftsmen
ICI Montreuil	2012	Montreuil (Paris region)	First makerspace created in Make ICI network, dedicated to craftsmen and creative industries
Cité de l'économie et des métiers de demain	2019	Montpellier	Innovation space dedicated to the management of open innovation projects for the elaboration of urban services based on deeptech, green tech, and digital technologies
I-Care Cluster	2009	Lyon	Regional cluster in Rhone Alps region on med techs and e-healthcare with a living lab
Hacking Health	2014	Strasbourg	Regional chapter of the international network Hacking Health, supported by a local firm specialised in healthcare. Hacking Health supports innovation in healthcare with user-centric approaches, and the organisation of hackathon-type events.

active in healthcare ecosystems. This research is based on 24 interviews and six visits to open labs between 2017 and 2019.

This chapter contains five sections. The first one introduces the Triple and Quadruple Helix approaches as innovation paradigms. Section 10.2 investigates how open labs operate as "boundary spaces" in the Triple Helix. Section 10.3 explores the contributions made by end users and civil society operated in the Quadruple Helix thanks to open labs. Section 10.4 discusses the respective contributions made by open labs to the dynamics of the Triple and Quadruple Helixes. A conclusion (Section 10.5) follows.

10.1 Triple and Quadruple Helix as innovation paradigms

With the open innovation paradigm, Chesbrough (2003) explains that public and private actors can no longer rely on their own internal resources to innovate. The knowledge base required for innovation is more distributed than ever among all these actors; it must combine internal and external assets to create new products and services. Open innovation requires the consideration of the dynamics of ecosystems and of communities to develop strategies. However, the innovation paradigm based on openness requires further characterisations because organisational drivers are not easy to identify, especially at ecosystem or territorial level. New roles emerge in this paradigm. The governance of innovation is transformed.

Gibbons (2000) complements the open innovation paradigm at territorial level: he explains how public and private actors deploy new processes and collective practices in ecosystems of innovation. With this approach, he also goes beyond the shortcomings of the linear model of innovation to explain how to consider new technological and societal challenges. This approach is coined as the Mode 2 model of innovation. It elaborates on non-linear processes and shows new ways of working for academics, firms, and policymakers, and interactions between them. Mode 2 is not necessarily disconnected from science and techno-pushed discoveries, but its starting point lies in societal concerns and anticipations about the adoption of innovation. Handling these aspects requires a multidisciplinary approach. It also requires an ability to systematically anticipate the accountability of promoters of innovation, most notably in relation to externalities.

Carayannis and Campbell (2009) are concerned with the generalisation of the open innovation framework. They expand the analysis because they consider that Mode 2 does not take into account the users' involvement enough during the innovation processes. With their shift from Mode 2 to Mode 3, they broaden the scope of multidisciplinary approaches suited to understanding interactions and ways of working with all categories of stakeholders. They also foster transdisciplinary thinking and problem-solving approaches. Problems are addressed by walking in the end users' shoes. End users' contributions become central to the development of new ideas, concepts, and prototypes (Von Hippel, 2005; Lettl et al., 2006).

Zooming out from Modes 2 and 3, it is easy to point out common features. These two modes of innovation management imply the adoption of new collaborative designs based on a multiplicity of partners. They also mandate the constant search for the alignment between various perspectives promoted by public and private actors. At territorial level, there is a growing tendency for the governance of innovation to be driven by collective strategies and for communities to attract an ever-growing heterogeneity of stakeholders.

Modes 2 and 3 of innovation challenge policymakers in several respects. Etzkowitz and Leydesdorff (1997, 2000) and Etzkowitz and Zhou (2017) consider that these models imply the development of the Triple and Quadruple Helix governance modes of innovation. The Triple Helix focuses on strong and virtuous ties between scientific institutions, firms, and policymakers for the production, diffusion, and exploitation of new knowledge. Each institutional sphere must be able not only to perform its own functions but also to take responsibility for some aspects of the other Triple Helix institutional spheres' traditional contributions (Etzkowitz, 2003; Etzkowitz and Leydesdorff, 2000; Hasche, Höglund and Linton, 2020). Boundaries between each institutional sphere tend to blur and the underlying processes contribute to the development of new regional knowledge architectures (Heraud, 2017).

Caryannis and Cambpell (2009) consider that the emergence of Mode 3 implies that innovation policy should shift from Triple Helix to Quadruple Helix innovation governance schemes. Collaboration now expands from three institutional spheres to encompass civil society, understood in a large sense: citizens, artists, media, etc. Challenges arise in relation to user-centric and societal perspectives wherever the Triple Helix institutional spheres are involved (Hoglund and Linton, 2018). Interactions in the Quadruple Helix require the development of new communities based on a high degree of trust between actors (Carayannis and Rakhmatullin, 2014).

The importance of the reference to Triple and Quadruple Helix mechanisms has already been identified for smart cities and healthcare systems (Leydesdorff and Deakin, 2011; Hoglund and Linton, 2018). The links between pharmaceutical and technology-intensive industries, academics, research hospitals, and policymakers have always been considered as key dimensions to encourage innovation in a techno-/science-push framework. However, the deployment of new technological opportunities and the materialisation of innovation is inseparable from new collective strategies between scientists, private firms, and policymakers (Merindol, Le Chaffotec and Versailles, 2022). Creating virtuous interactions between public and private actors is also a central issue in smart city programmes. In both areas, the design of innovation-related policies progressively shifts from the Triple to the Quadruple Helix approach. There are two reasons for this evolution: first, consider civil society needs in parallel with science-/techno-push approaches and, second, promote initiatives fostering coproduction with end users (Paskaleva et al., 2021). In the real world, the words "civil society" and "end users" cover lots of different categories of people. In many cases, students and employees can be counted as end users of new technologies. Their visions, needs, and emotions can be considered as relevant inputs and collected with creative protocols such as design thinking. For smart cities, users are also journalists, citizens, artists, and people from the entertainment industry. In healthcare ecosystems, "users" cover many different categories: practitioners (physicians, nurses, and any professional in charge of medical care), patients, their relatives and anyone supporting them (Schiavone, 2020; Bjørkquist et al., 2015). From a practical point of view, the challenge is to find relevant ways to associate all categories of users to innovation processes and consider all their points of view.

Managing the Triple and Quadruple Helix models of innovation remains a complex venture. Brokers are required, who play a central role to foster and feed new connections and to enhance the fluidity of exchanges among actors present in the innovation process. Such brokers are both individuals and organisations. In the Triple and Quadruple Helix perspectives, organisations considered as brokers are mainly called hybrid organisations. Collective brokers operate in relation to various institutional spheres, or at the intersection between them (Champenois and Etzkowitz, 2018). Individual brokers are usually identified as boundary spanners (Lundberg, 2013). They contribute to translate meaning, and to create common understanding, for people coming from different backgrounds. Boundary spanners and hybrid organisations make the integration and the combination of various elements possible. They manage spaces of interactions suited to break silos and to orchestrate new collaborative designs for innovation (Etzkowitz, 2003; Champenois and Etzkowitz, 2018). All these brokers intervene in a multilevel perspective to create the conditions of individual contributions to innovation processes.

Sarpong et al. (2017) acknowledge that the adoption of the Triple Helix model implies a twofold transformation of practices: an increase in the permeability between public and private organisations, and the development of collective entrepreneurial behaviours to sense new markets and societal needs before generating the associated innovation patterns. Dzisah and Etzkowitz (2008) emphasise the necessity to generate fluidity and to improve the circulation of ideas. They focus on the importance of developing reciprocal behaviours to encourage new contributions in a reflexive way. Brokers contribute to the installation and operation of such fluidity.

10.2 Open labs in the Triple Helix approach

The portfolios of services offered by open labs have been extensively discussed in Chapters 1 and 2. This section offers the opportunity to identify the diversity of their contributions to develop the governance of innovation based on the dynamics of the Triple Helix in the domains of smart cities or healthcare systems (10.2.1). It also illuminates roles enacted by open labs to develop new collective practices for innovation at the intersection between these three institutional spheres (10.2.2).

10.2.1 Contributions by open labs to the dynamics of the Triple Helix

Institutional spheres playing a strategic role as catalysts of the creation of open labs vary. In the list presented in Table 10.1, four open labs are driven by local policymakers, three open labs are initiated by private companies, while only one is led by an academic institution.

When innovation governance refers to the Triple Helix, Ranga and Etzkowitz (2013: 242) explain that institutions become catalysts of innovation-related collaborative projects when they elaborate on "innovation organisers": individuals "who typically occupy a key institutional position, enunciate a vision for knowledge-based development and have sufficient respect and authority to exercise convening power to bring the leadership of the institutional spheres together".

The cases show that "innovation organisers" are driven by concerns varying with the institutional sphere these individuals come from. When open labs are steered by local policymakers, "innovation organisers" are politicians exhibiting a legitimacy built with the development of the city or in the healthcare ecosystem. They show concrete personal records in these domains and have demonstrated a capacity to assemble resources and efforts at territorial level. They are acknowledged as individuals able to unite public and private efforts around them. Innovation organisers face a challenge: convince the other institutional spheres to be part of the open lab project to accelerate innovation for the benefit of the whole ecosystem. When local policymakers steer the creation of open labs and propose principles for their organisation, they progressively convince private companies and universities (the two other pillars in the Triple Helix) to play an active role in the development of these activities, and most notably to participate in experiments operated in the open lab, to creativity events, and to different categories of workshops and conferences. These "innovation organisers" also contribute to the definition of new services.

This evolution is typically instantiated by TUBA, located in Lyon. Local policymakers were committed to launching new dynamics for the (digital) transformation of the Lyon urban area. The president of Lyon metropolitan government was conscious that the digitalisation of the economy and of the ecosystem was making it mandatory to think differently about urban public and private services. He supported the development of TUBA open lab to accelerate these transformations, with a twin mission to support experimentation of digital technologies in urban areas and handle the associated data. He was at the initiative of the creation of the open lab and influenced its missions with a clear vision of services to be delivered to the community. He did not want the administration to operate TUBA, but public subsidies were important at the beginning of its story. This approach convinced large companies and universities to join this journey. Boosting collaboration with large companies and universities was easy in the ecosystem because the dynamics of the Triple Helix were well established in the territory before the creation of TUBA. This interaction concretised with the improvement of urban services.

In Montpellier, the regional agency in charge of economic development and local politicians (embedded in the city, with national elective mandates) are the origins of the initiative. They intend to translate the innovation strategy defined for the region and the metropolitan area in healthcare, green technologies, "big data" management, and quantum computing into an attractive hub where start-ups, large established companies, and basic research labs work together at the interaction between experimentation, open innovation, and foresight about education. The ultimate strategic intent is to run experimentation with other stakeholders, use the outcomes of innovation projects to develop prospective actions and grasp the evolution of jobs and competencies for the future, and prepare the associated education programmes. This is the reason why the open lab has the name "City of economy and professions of tomorrow" (Cité de l'économie et des métiers de demain). Local policymakers behave here as innovation organisers. The initial effort to give momentum to this open lab requires significant resources because traditional contributors to the Triple Helix are not used to working together on these issues. They are also surrounded with several other successful open lab initiatives attracting start-ups and large companies that are not already installed in Montpellier. This means that competition between open labs will require companies to operate strategic choices because they cannot be present everywhere at the same time.

In Strasbourg, the creation of Plages Digitales (literally "Digital Beaches", two coworking spaces) and "La Fabrique" ("The Factory", a makerspace) are also driven by local policymakers. Their goal is to create new spaces to encourage digital transformation and the diffusion of makers and entrepreneurial mindsets in the ecosystem. Local policymakers acknowledge that they do not have the appropriate competences to operate urban transformation with digital technologies and design these open labs. That is the reason why they decided to subsidise a non-profit organisation called "Alsace Digitale" driven by local digital entrepreneurs to promote digital transformation on the territory, that in turn launched the Plages Digitales and La Fabrique initiatives. Entrepreneurs in Alsace Digitale facilitate the development of the associated communities. Policymakers also convinced other actors such as (private) established companies and academic institutions to collaborate to projects steered by the entrepreneurs hosted in these open labs. Large companies have donated equipment and tools for the La Fabrique makerspace. The University of Strasbourg regularly sends students to hackathons run by the Plages Digitales. Even if contributions by the University of Strasbourg and by large companies are limited to specific actions, these open labs offer the opportunity to create new links between public and private actors to manage the (digital) transformation of the city. These initiatives offer the opportunity to install the dynamics of the Triple Helix on this specific topic.

When the open lab project is driven by private firms (or entrepreneurs) or by an academic institution, collaboration modalities in the Triple Helix follow different patterns. When people coming from academic institutions or private companies act as innovation organisers to deploy an open lab, their original "institution" remains the focal player to develop the open lab's activities. They look for other actors to contribute to the dynamics of innovation and become partners in the different activities. They always lead the open lab's activities.

ICM is a research organisation active in healthcare, operating with an incubator and a living lab. Its goal is to promote an innovative ecosystem specialised in diseases of the spinal cord and of the brain. ICM expects to attract students, firms, entrepreneurs, and healthcare practitioners to work with scientists on the development of new innovative solutions and to facilitate a rapid transfer of academic inventions to private firms. For ICM, the challenge is to convince a diversity of actors to partner at different steps of the development of innovative solutions. The dynamics of collaboration were easy to install because of ICM's reputation of scientific excellence, at two levels: personal (researchers, innovation organisers) and organisational (team effectiveness, success with international programmes and grants). Such aspects contribute to build legitimacy and create a climate of trust around the community and the open lab, thus raising its attractiveness and visibility as focal actor in this research domain. It is to be noted that the dynamics of the Triple Helix were present around ICM before the creation of ICM's open labs: private companies, the hospital (Paris Pitié Salpêtrière University Hospital) and policymakers have been working together since the installation of ICM as Excellence Institute (IHU, in French terminology) to accelerate innovation transfer mechanisms. The creation of ICM's open lab has offered the possibility of expanding interactions among these institutional spheres.

The role of private firms in the development of open labs and as drivers of new collaborations in the Triple Helix varies with the size of the company at the origin of the project, and with the motivation of entrepreneurs promoting the project. thecamp perfectly illustrates the case of a successful entrepreneur with a mission to contribute to social transformation, who intended to search for disruptive solutions in relation to climate change and sustainability. The open lab has facilities between Aix-en-Provence and Marseilles Provence airport, in the south of France. The entrepreneur at the origin of project (see Chapter 2) managed to convince large companies from different sectors and local policymakers to partner in the project, but the governance is clearly in favour of companies. The entrepreneurial logic remains at the centre of the management of the open lab. At thecamp, the managerial team has an entrepreneurial record; it has installed co-creation activities and facilitates experimentation projects. Large companies, local policymakers, and universities contribute to the projects, but the open lab remains fragile, and the business model takes time to stabilise (cf. Chapter 2), thus requiring adaptations after several months of closure at the beginning of the Covid pandemic. Contributions by large firms are important and were present in the portfolio of activities from the earliest stages of the open lab design, but the implication of universities and basic research remains low. It is difficult to assess whether this situation represents the consequence of the design itself, an outcome of its governance design, or a matter of the Triple Helix dynamics. The interaction with local policymakers (metropolitan and regional governments) represents a difficult journey because thecamp does not want to take sides in the competition between the cities of Marseilles and Aix-en-Provence; it has decided to adopt a position of neutrality between all public actors that are engaged in a fierce battle of influence against each other in this ecosystem. The contribution by local universities and tertiary education institutions remains low and limited to specific programmes, mainly due to internal reasons in relation to governance issues and the implementation of internal reforms, that make it difficult to commit as strategic partner.

Hacking Health represents an original initiative fostering the creation of new links between public and private actors of the Triple Helix (see Chapters 4 and 5 in this book). The Hacking Health chapter installed in Strasbourg is managed by a local firm specialised in innovation in healthcare located in the same city. Each year, it organises several "small" events (such as seminars about design thinking, or conferences about new topics relevant for healthcare) to prepare for the annual hackathon gathering between 300 and 750 people. These activities are based on a collective strategy orchestrated by the local Hacking Health chapter: Strasbourg University, local policymakers, and France Biovalley innovation cluster contribute to the organisation and to the animation of the various activities surrounding the hackathon. All activities, and most notably the large final event, contribute to break organisational and institutional silos and to create new interactions for innovation in the local healthcare ecosystems.

10.2.2 Open labs as "boundary spaces" in the Triple Helix

The Triple Helix approach requires the development of various spaces of interactions between academic institutions, private firms, and policymakers. Ranga and Etzkowitz (2013) provide a taxonomy of spaces that contribute to the integration of various elements proposed by the institutional spheres of the Triple Helix.

- The "knowledge space" makes it possible to intensify interactions between stakeholders, and to produce new knowledge.
- The "consensus space" empowers contributors to the Triple Helix to create a consensus between the institutional spheres, most notably about priorities for innovation policy.
- The "innovation space" contributes to find new solutions based on the integration of competences and assets of the Triple Helix's institutional spheres.

Champenois and Etzkowitz (2018) propose to regroup these perspectives under the term of "boundary space" to illuminate a common feature: they all contribute to the building of bridges over the cognitive, spatial, and organisational boundaries existing between the institutional spheres of the Triple Helix.

Open labs dedicated to smart cities and healthcare systems perfectly exemplify these boundary spaces, and boundary-spanning activities. They elaborate on both cognitive and physical dimensions. In these spaces, representatives of the three institutional spheres have interactions, and collaborate.

Open labs are defined as "*neutral spaces*" by many of the interviewees. In open labs, people from various institutional spheres negotiate new projects and their contributions. The catalyst role is enacted by managers and facilitators present in each open lab's team: these individuals create the conditions for collective action, contribute to connect people together, to align interests and to translate meanings from an institutional sphere to the other(s). Individuals in the open lab teams are sometimes academics and researchers, people from the industry, or from public institutions. They collectively act as "boundary spanners" and animate the various spaces of interactions of the "boundary spaces" present in the Triple Helix.

The cases about healthcare and smart cities in this chapter show that open labs do not equally contribute to the development of the three subcomponents of the "boundary spaces" (consensus, innovation, and knowledge spaces) and do not equally focus on their promotion.

In the sample, TUBA and thecamp promote all three aspects: consensus, knowledge, and innovation spaces, even though their respective originalities do not follow the same sequences. TUBA's first function is to create a consensus among partners about how to collectively develop new services to enhance the ecosystem with smart city services. An executive from a large company explains: "TUBA is the right place to define what we want to do for smart cities with various partners. This place is politically neutral enough to consider it the right place to have open discussions and collectively define what to do". As TUBA also organises meetings and brainstorming sessions about best practices and data management about smart cities between people coming from each part of the Triple Helix, the open lab should also be considered as a knowledge space. TUBA's physical space is deliberately designed and configured to foster such interactions. Even its location close to Lyon Part Dieu transportation hub (and station) is carefully selected to serve this purpose. A contributor from Lyon University explains that they organise their meetings there "when they want to have interactions about solutions, best practices, and rules to be applied. TUBA is convenient because of its various facilities: brainstorming and creativity room, canteen, open office, etc.". Last, but not least, TUBA's managerial team also supports the collaborative process and the installation of experiments of new digital solutions about smart cities, thus deserving the qualification of innovation space.

The function of consensus space also comes first in the strategy of thecamp, but it originates in specific expectations and initiatives led by the industry. Public and private actors then negotiate contributions to deploy the open lab's activities, create knowledge together and co-create new solutions. The initiative belongs to large companies operating in different sectors that populate the different governance boards of thecamp, and purchase bulk services during the negotiation of activities for each fiscal year. Their interaction, and the associated co-creation processes also mobilising Triple Helix actors that do not belong to the network of strategic partners, characterise the consensus space. Two main tasks can therefore be identified for the managerial team: first, at strategic level, facilitate meetings defining a consensus about priorities for the yearly agenda of activities in the open lab (main topics, list of experiments, list of services, etc.); second, at operational level, facilitate meetings and activities to generate a consensus about experiments (methods, protocols, returns on experience, conclusions, lessons learned, etc.). As explained by a representative from a large company that is among thecamp's strategic partners, this open lab is "*a place where large companies that have no direct working connections can define a collective strategy to address sustainability issues and identify their respective forthcoming developments*". Even though thecamp has a vested interest in delivering effective services to the large companies committed as strategic partners, it also serves a knowledge and innovation space to discuss digitalisation and societal transformation with academics and policymakers. The innovation space function is not only concretised with the discussion of protocols, methods and returns on experience. Facilities are also used to experiment, as it is typically instantiated with the experiment about autonomous vehicles developed on private roads between its facilities and Aix-en-Provence TGV (high speed train) station. thecamp therefore deserves the qualifier of innovation space because it hosts prototyping co-creation activities with scientists from different countries, entrepreneurs, engineers, and employees from large companies, and policymakers.

In Montpellier, the "City of economy and professions of tomorrow" (Cité de l'économie et des métiers de demain) has been designed to operate as a sort of copy-paste of Tuba and thecamp, two open labs that explicitly inspired its strategy. As it started its activities in late 2019, only the function of knowledge space could be installed with a series of events and meetings. However, the strategic plans are designed to expand to two additional directions. First, experimentation and the hosting of start-ups committed to the development of innovative solutions in relation to the regional and metropolitan research strategy, thus adhering to the role of innovation space. Second, facilitation of networks of experts and communities in relation with these specialisation domains and support to the installation of experimentation protocols (including at the level of the analysis of data generated during experiments), thus acting as a consensus space and as the neutral actor making the emergence of consensus possible.

ICM provides an illustration of the development of knowledge and innovation spaces in relation to healthcare: it offers the opportunity for entrepreneurs, large companies, professionals, and scientists to exchange about brain and spinal cord diseases. Their colocation in Pitié Salpêtrière University Hospital makes it much easier to organise interactions and work together about best practices and new solutions. ICM facilitates seminars and experiments about digital applications in the hospital. They also organise collaborations between scientists and start-ups together using research and technological platforms available in ICM's facilities.

Whatever the rationales leading to the installation of the dynamics of the Triple Helix in an ecosystem, open labs represent the right place to enact open innovation and entrepreneurial ventures. Teams managing each open lab encourage reciprocal behaviours and promote collaboration. In TUBA, firms accepted to partner without previously defining the nature of services delivered by the open lab and without identifying the nature of their contributions:

> We [i.e., large companies] agreed to join the project and to collectively participate to this innovation journey. We noticed opportunities to develop projects about smart cities and work about their deployment but we do not have any idea about the final results of TUBA's project.
>
> (Representative from a large company)

Table 10.2 provides an overview of collaboration practices for the open labs in the sample. It also displays the nature of the respective innovation organisers and of components of the boundary space (consensus, knowledge, innovation spaces).

10.3 Open labs extending the dynamics of innovation to the Quadruple Helix

The Quadruple Helix approach encourages active contributions by users, citizens, artists, and media in the development of new innovative projects. The cases show how open labs contribute to cope with big societal challenges and elaborate on contributions by civil society to identify and appraise potential solutions. Carayannis and Rakhmatullin (2014) highlight that the Quadruple Helix is not only about involving end users and citizens in exploration projects driven by universities, firms, and policymakers. The Quadruple Helix approach also deals with actors from civil society who suggest new innovative solutions. The extension of collaboration to civil society represents a big challenge for universities, firms, and policymakers. This implies an environment suited to encapsulating a myriad of relationships and visions (Hoglund and Linton, 2018). In the case of smart cities and healthcare ecosystems, the variety of actors is potentially very large. Encouraging the Quadruple Helix model also implies a change in the culture of innovation.

This chapter reveals how open labs running smart cities and healthcare innovation projects provide several initiatives to operate in the framework of the Quadruple Helix approach. Different categories of end users and stakeholders become the centre of the exploration process. Open labs demonstrate some creativity to accommodate these people in the development of new projects. In many cases, end users and civil society actively contribute to problem definition and resolution, but the cases also show how difficult it is to transform them into active contributors to the innovation processes.

Managers of open labs interviewed during data collection point out the necessity of developing a user centric approach of innovation in the domains of

Table 10.2 Open labs in the dynamics of the Triple Helix

Open labs	*"Innovation organiser"*	*Components of "boundary space"*			*Collaboration practices*
		Consensus	*Knowledge*	*Innovation*	
Tuba	Local policymakers	☑	☑	☑	Cooperation between large companies, local universities, and entrepreneurs about digital urban services: collective definition and experimentation
Plages Digitales La Fabrique	Local policymakers	☐	☑	☑	Collaboration between entrepreneurs to propose new initiatives supporting the digitalisation of the ecosystem to local policymakers and develop a user-centric culture of innovation
ICM Incubator cLLAPS	Research institution	☐	☑	☑	Organisation of hackathons and collaboration projects: healthcare practitioners, students from design, engineering, and business schools, start-ups, etc. The aim is to bring digital applications into the hospital
thecamp	Entrepreneur	☑	☑	☑	Collaboration between artists, large companies partnering with thecamp, start-ups and academics to support the transformation of large established firms and their creative capabilities
ICI Montreuil	Entrepreneur		☑	☑	Collaboration with local policymakers to support craftsmen and small firms specialised in the creative industries
Make ICI	Entrepreneur	☐	☐	☑	Ad hoc collaborations in each town to rejuvenate industrial wastelands located in urban areas with partners in industry (e.g., Bouygues in Marseilles) and foster the installation of craftsmen or companies
Cité de l'économie et des métiers de demain	Local policymakers	(P)	☑	(P)	Currently only keynotes organised for firms, start-ups, end users, citizens, and academics. *Aspects marked with (P) are postponed after the pandemic*
I-Care Cluster	Local policymakers	☐	☑	☑	Collaboration between firms, academics, and healthcare institutions (most notably foster homes) to install mini-labs and develop special events on innovation for practitioners
Hacking Health Strasbourg	Entrepreneur	☐	☑	☑	Co-funding and facilitation of hackathons on e-healthcare with local policymakers, companies, and academics

healthcare and smart cities because the acceptability of innovation represents an important challenge. This issue implies an acknowledgement of the importance of empathetic behaviours and the necessity of combining approaches considering the different aspects of innovation adoption (Paskaleva et al., 2021), most notably rational arguments about personal benefits and local advantages, the complexity of proposed solutions, strategies towards trialability, and observability, but also more subjective aspects in relation with emotions and aesthetics. Open labs develop original initiatives to manage direct contributions by civil society and the end users to new projects with experiments and trial-and-error processes about concrete scenarios and adoption cases.

This research identifies three main contributions by end users and civil society in the development of collaborative projects in open labs. These contributions imply different degrees of commitment at various phases of the innovation processes. The subsequent subsections illustrate and analyse the reference to insights proposed by the civil society (10.3.1), their role as co-creators in innovation projects (10.3.2), and their specificities as actors of experimentation (10.3.3). The three components of the "boundary spaces" (knowledge space, consensus space, innovation space) resonate in different ways with the implication of civil society in the development of smart cities and healthcare systems. Table 10.3 wraps up on the different contributions identified in each open lab of the sample.

10.3.1 Open labs promote civil society's insights in the creative process

Open labs develop several initiatives to consider direct suggestions emanating from end users and civil society. Their challenge is to bring new ideas and insights at the kick-off of projects managed by large companies, local policymakers, or universities. In the sample considered in this chapter, three open labs develop explicit initiatives to bring insights in creative processes in line with the Quadruple Helix approach: thecamp, ICM, and TUBA.

In projects about innovation on societal issues (e.g., autonomous vehicles, the evolution of cities, the future of housing), thecamp launched different initiatives

Table 10.3 Open labs in the dynamics of the Quadruple Helix

Open labs	*QH*	*Contributions by the civil society*
TUBA	☑	New insights and experimentation
Plages Digitales/La Fabrique	☑	Co-creation (La Fabrique/makerspace)
ICM Incubator/cLLAPS	☑	Experimentation and co-creation
thecamp	☑	New insights, experimentation, and co-creation
M ICI/ICI Montreuil	☑	Experimentation and co-creation
Cité de l'Économie	(P)	(*Planned, not yet implemented*)
I-Care Cluster	☑	Experimentation
Hacking Health	☑	Experimentation and co-creation

to consider points of view coming from civil society. It also facilitates seminars where children, teenagers, and artists contribute to prospective actions for companies. During creativity seminars, teenagers and children react to videos, models, and drawings picturing architectural projects produced by architectural firms. They propose new ideas, and these seminars contribute to modify the vision of future introduced by these companies. A manager at thecamp explains:

> Collaboration between universities, companies, entrepreneurs, and public policymakers is not enough to develop smart city projects. [...] We want to illuminate these collaborative projects with new insights coming from various protagonists. [...] It represents one of the conditions for thinking out of the box about the future of the society.

ICM living lab also offers several instances of how to deal with visions, emotions, and returns on experience collected from healthcare professionals in the process of innovation. The living lab is located in the largest university hospital centre in Paris. cLLAPS knows healthcare professionals very well and has developed trusted relations with surgeons, medical doctors (MDs), and nurses. The living lab hosts short seminars with them on a regular basis to develop informal exchanges. According to ICM cLLAPS's manager:

> This creates a clear vision of the culture, practices, and challenges for the hospital. This shows what is really appraised as important by professionals in the hospital. Creating a climate of trust and the fluidity of exchanges with healthcare professional takes time. Professionals are trapped in daily routines. They do not have much time to spend about innovation. [...] It is a cultural problem. Talking with a critical eye about how to improve their professional practices is new and difficult.

The challenge is to take advantage of ideas stemming from professionals who work in the hospital to innovate "from the inside", thanks to comments about what works well and what does not.

At TUBA, activities develop around the conception and prototyping of new services for the Lyon metropolitan area. Citizens are the end users of services to be installed in the future, and the targets of public policies. Their participation to innovation processes is a key concern in this open lab. In 2017, TUBA launched a community café ("café communautaire") in the same premises as the open lab, with the aim of attracting people at random, just because they want to have a drink or a snack, and the hope that they would eventually make small or more consistent contributions to the ongoing projects. It should be noted here that the entrance of the open lab was then located on the same square as one of the main train stations in Lyon, with lots of (national and international) travellers and even more commuters walking past the door every day. The strategic objective was twofold: first, make sure that these people could find a place to voice their points of view; second, randomise the access to "normal" people, and avoid selection biases. The location of the café was already attractive to people representing the traditional stakeholders of the Triple Helix, who soon started to schedule meetings for the convenience of it. The café also gave the opportunity for all "normal" people (travellers or commuters) to just sit there and work for a moment or for several hours. They also had opportunities to interact with start-uppers or experts present for the projects. They could ask questions, attend seminars on TUBA's topics, or register to take part in later experiments. TUBA's team decided not to continue the "café communautaire" after 18 months. The managing director then explained: "It was very complex to create an impulsion for the participation of citizens. We spent a lot of energy in terms of animation, but only for scarce results". The challenge was to convert high-level discussions into actual contributions about the transformation of the city and of its public services. Most people just walking by and entering "at random" were often seeing TUBA as one of the pubs on the square, albeit an original one, and not seriously considering any commitment to TUBA's other activities. This low conversion rate made the "café" some kind of sunk cost. This case illustrates that generating active contributions by civil society remains a tricky endeavour.

10.3.2 Civil society as co-creator in open labs

Several open labs of the sample considered in this chapter have actual experience in associating civil society with the development of new projects as co-creators (see Table 10.3). The physical space installed by these open labs represents an appropriate arena for such kinds of initiatives.

Relying on its initial success, the team at the origin of ICI Montreuil decided to launch a network of open labs in various French cities under the "Make ICI" label (see Chapter 1). The common ground for this network is twofold: first, develop resources with a motivation at local level, yet pool them at network level; second, commit to the rejuvenation of industrial wastelands in urban districts. Each open lab in the ICI network installs specialised activities where its contribution makes sense because of a specific tradition in a local ecosystem and of possible interactions with other firms. Each "ICI" branded open lab therefore proposes various arrangements with local economic actors around a boundary space meeting this tradition of distinctive local competencies; it tailors the service portfolio to meet the local needs. This leads to different types of machines in "ICI" open labs, thus progressively building complementarities at network level. As Make ICI CEO and ICI Montreuil's founder explains:

> Make ICI offers opportunities for local policymakers to reindustrialise the centre of their city, and support craftsmen by designing buildings where they can transfer their businesses. In this framework, we support communities of craftsmen who find opportunities to bring their original contribution to the development of smart cities. Make ICI and craftsmen become key players of urban planning.

Such planning does not occur in a top-down manner and the boundary space is conceived as an enabler. As in Montreuil, the makerspace is understood as the initial investment that will support the development of the dynamics of each ecosystem. Each open lab becomes a sort of focal actor making the conditions for rejuvenation possible. It leaves it to civil society to build its own trajectory and do something with it. In the Make ICI network, each open lab represents an original way to include civil society in the elaboration of new urban designs and to encourage new collaboration models with local policymakers.

The same logic applies to hackathons operated by Hacking Health Strasbourg chapter. Healthcare practitioners, most notably physicians and nurses, find the opportunity to contribute to the elaboration of digital solutions meeting their own needs there. The yearly hackathon serves as a sort of boundary space where all sorts of volunteers meet for three or four days to develop the initial phases of innovation projects with creativity methods. To ensure the appropriate volume of contributors, the Hacking Health chapter coordinates with local universities and ensure the presence of students from different specialisations, at different levels (bachelor's or master's curricula). Entrepreneurs interested in the topics, or already having a mock-up available, will also join activities. The end of each hackathon represents a starting point for increased specialisation in the teams, for instance to further experiment in another environment, or develop improved prototypes and commit to a more structured innovation project. Each hackathon represents an instance of boundary space creating the conditions for breaking silos, generating unexpected encounters, and developing something concrete with collaborations that would not have otherwise emerged. The retrospective feedback collected from students and volunteers from the civil society is always the same: none of them would have imagined that they were able to make valuable contributions to creativity or innovation processes, and they are all willing to support the development of "their" project in the future with additional commitments.

These cases both illustrate the relevance of installing opportunities for civil society to contribute to innovation projects. When dealing with complex issues such as the rejuvenation of industrial wastelands, or healthcare-related innovation, the contributions of civil society materialise when a boundary space creates the appropriate conditions to host efforts and initial collaboration and operates with the appropriate facilitation methods.

10.3.3 Civil society as actor of experimentation in open labs

Open labs offer an opportunity to organise experiments in a different way. The characteristic presentations of end users' and citizens' contributions to experiments make them guinea pigs testing the different solutions in real life environments like they would do in living labs. However, projects are not limited to observing actual people experimenting. Here an originality arises when discussing new solutions for healthcare systems and smart cities. Citizens, patients, healthcare practitioners provide feedback about new solutions as actual experts of their own diseases, or of their own living environments. In open labs specialising in these topics, end users and citizens have an opportunity to contribute to the design of experiments, to appraise data-collection processes, and to make sense of data alongside the other "experts". These aspects go beyond traditional aspects computed for "experimentation".

This chapter has already mentioned TUBA several times. TUBA's team organises actual areas in the urban space for experimentation and convinces citizens to participate to these activities. They do not need an actual living lab because the urban environment itself represents their testing ground. However, they acknowledge that the mobilisation of actual citizens remains a challenge while framing experimentation protocols with start-ups, large companies, and academics is much easier. TUBA knows how to bring in specialised expertise such as designers, sociologists, and ethnography experts for the evaluation of experiment results. To make the actual contributions to experimentation easier, TUBA has also set up a coworking space with low prices, where it hosts freelancers and entrepreneurs. In return, these coworkers are required to participate in experiments and replace hard-to-find volunteering citizens. This initiative replaces the "café communautaire" mentioned earlier. However, contributions do not cover the same reality because the coworking space attracts people who already work on urban issues, such as architects, or specialists in digital solutions for mobility.

The case of ICM's cLLAPS living lab is interesting to describe how civil society contributes to experiments in the domain of healthcare, with the originality of interactions between patients, families, and practitioners under very high safety constraints. Activities elaborate on actual hospital rooms and actual patients, either in ICM's research facilities, or in the hospital itself. Here lies the importance of locating ICM and its living lab in Pitié Salpêtrière University Hospital in Paris (the largest French hospital). Physical proximity makes day-to-day interactions easy. It eases negotiations for the installation of experiments and, also, provides visibility and legitimacy to cLLAPS's activities. The living lab's team closely works with doctors, nurses, all categories of healthcare practitioners to test new solutions; it `collects feedback from patients and families. With experimentation activities inside actual facilities, and inside actual medical care pathways, activities accelerate the innovation process and prepare for more effective diffusion of innovation. There is also no issue in transitioning from lab-based experiments to real life because testing complies with hospital standards and is operated by actual people: they are no experts in experimentation but have an explicit expertise with their jobs. These aspects combine to explain a greater transition and a smoother diffusion for digital applications and innovations because they are directly adapted to healthcare needs.

All these elements explain why experimenting with users represents a very nice option to hasten innovation processes: it targets actual problems and proposes solutions that are more readily prepared for diffusion. From this respect, open labs propose interesting solutions because they anchor experiments within a context of feasibility, legitimacy, and trust. This is an important contribution of "boundary spaces". At ICM's cLLAPS living lab, contributors directly grasp forthcoming benefits. However, TUBA illustrates that experimenting with civil society faces a major challenge: how is it possible to incentivise actual people to contribute to innovation projects?

10.4 Discussion: open labs as catalysts in the dynamics of the Triple and Quadruple Helix

Open Labs considered in this chapter represent "boundary spaces". All are located at the intersection between the various institutional spheres of the Triple and/or Quadruple Helix. They contribute to the evolution of mindsets and to the dissemination of a culture of innovation by supporting the dynamics of the Triple and Quadruple Helix. This chapter makes four main contributions.

First, this parallel analysis of cases in healthcare and smart city ecosystems offers the opportunity to better understand the complex interactions between local policymakers, industry, and the actors of basic (and applied) research to design open labs. Activities may originate in any Helix, but effective interactions always require the development of joint efforts. Depending on the institutional sphere that plays the role of "innovation organiser", contributions by the Triple Helix institutional spheres will vary and will not have the same levels of importance when deploying the open lab's services. The cases show that the academic world will eventually lead open innovation projects in existing open labs. However, in France, we observe that most open labs are driven by industry and by policymakers. Is this a French specificity? Our research into healthcare and smart cities could not identify any case installed by an academic institution unless living labs created and operated by university hospitals (see Chapter 4) qualify in this category. Considering the specificities of research hospitals in France, we would not characterise their living labs as independent ventures.

Second, this research also identifies the importance of boundary conditions in the dynamics of collaboration developed around open labs. This is most notably the case with the nature of interactions predating the activities of open labs in the Triple Helix. This research suggests that the ramp-up of activities around an open lab is easier when the dynamics of the Triple Helix already exist at the local level. Gaining momentum will then relate to the dynamics of the Triple/Quadruple Helix itself and will be much easier in the associated open lab if the actors of basic research, policymakers, and industry have already demonstrated the relevance of collaboration. They are therefore interested in developing their collaboration further; they host joint efforts in an open lab because it makes the governance of projects more effective. In this context, open labs become the pragmatic by-products supporting the consolidation and expansion of collaboration. Open labs will also eventually become the symbolic materialisation of such a win-win-win situation (like a sort of totem place). Open labs also offer the opportunity to create a neutral space to progressively develop new interactions between policymakers, companies, and academics.

This research also investigates other cases where interactions emerge with the creation of the open lab to follow the Triple/Quadruple Helix patterns. The open lab is then most often launched at the initiative of one single institutional Helix and becomes the basis for the elaboration of new collaboration patterns. In this context, open labs struggle to legitimise their presence beyond the sponsorship of one specific Helix. This process takes time. Tensions between Helixes will eventually make it difficult to scale-up in the ecosystem. The success of innovation projects and the explicit added value of the open lab both represent mandatory conditions to nurture legitimacy and trust in the competitive environment between the various institutional spheres.

The third point deals with ways of working prevailing in open labs. Contributors to innovation projects find a possibility to experiment with new collective and managerial practices in these interaction spaces. They directly experience a social learning cycle based on exchanges of concrete and uncodified knowledge that make it possible for all actors to collaborate and experiment together towards innovation. New justifications for the emergence of open labs emerge when these stakeholders develop the various spaces of interactions listed in Section 10.2.2. Exchanges and interactions in an open lab not only represent unique and non-replicable resources fostering the emergence of the consensus and innovation spaces, as in the resource-based view (RBV) of the firm. Where knowledge is unique and difficult to articulate and requires direct interactions about concrete problems and experiments between different categories of stakeholders, transactions about knowledge, joint interpretations of experiments, elaboration of consensus, and value creation through innovation have a strategic value. This diverges from the RBV approach and enters the knowledge-based theory (KBT) that was usually applied to firms, as identified by Ihrig and MacMillan (2013). The KBT approach is easy to transpose to open labs as "boundary spaces" to explain their strategic importance. The existence or evolution towards a knowledge space represents a justification for having an open lab, and for joining its activities.

The difference between the RBV and KBT approaches shows the explicit importance of open labs in Triple and Quadruple Helix environments. The interpretation of data made available with experimentation and the elaboration of consensus about data interpretation prove to be more complex (not automatically more difficult) in the framework of the Quadruple Helix because of the nature of topics addressed with actual end users and with the civil society. These elements explain why the dynamics of interactions and the ways of working inside the Helixes or at the intersection between them (basic research, policymakers, and industry, with or without civil society) lead to practical questions about project and operation management in open labs. There is no doubt that these aspects stress the importance of management teams in open labs, and of managerial processes. Interactions towards the elaboration of "boundary spaces" do not emerge at random. They must be managed at the level of the whole ecosystem with the other actors present in the Helixes. Explicit challenges exist for the design and the management of these spaces with collaboration and multi-partner projects.

An open question remains. What is the cooking method appropriate to articulate the subcomponents of "boundary spaces"? The cases of open labs considered in this chapter show *ex post* that the receipt elaborates on doses of knowledge space, consensus space, and innovation space. However, it is still difficult to characterise *ex ante* the sequence between these aspects and the appropriate ways of working enabling the emergence of effective solutions for these three spaces. Does the knowledge space come first? Cases considered in this chapter suggest that it all depends on the problem under consideration. The KBT approach eventually supports the primary importance of the knowledge space because of the difficulty at articulating knowledge assets inside the social learning cycles (Boisot, 1998) framing interactions, the elaboration of consensus, and the different aspects of project management applied to innovation projects. However, data collected about the open labs presented in this sample cannot lead to a generic analysis of the space that *drives* operations in an open lab, and its influence on the ecosystem. There is no "one size fits all" rule for the eventual logical or temporal precedence of knowledge, consensus, or innovation space against the other two.

The last contribution more precisely deals with the implementation of governance modes specific to the Quadruple Helix. This research stresses the originality of the role of the open labs when overseeing the association of civil society to innovation processes. Most open labs struggle to encourage the participation of civil society to innovation projects, but they can build tools to reach out to specific categories of citizens and build visibility (for instance with showrooms and the organisation of conferences, keynotes, etc.). Most of them also progressively build brands embodying the management of innovation in local ecosystems. These aspects go beyond the experimentation of new working modalities inside the Quadruple Helix governance model. This research identifies three types of involvement of the civil society during the innovation processes: as providers of insights in creative processes, as actors of experimentation, and as co-creators of the "boundary spaces". Open labs offer various channels to encourage the participation of the civil society to the development of new projects driven by universities, policymakers, and firms, but it remains difficult to have a definitive understanding of their direct contributions to the subcomponents of "boundary spaces". The cases illustrate for sure that civil society plays an important role with the two first types listed above. However, the cases also suggest different stages for its contribution to the co-creation of knowledge, consensus, and innovation spaces. Cases show that civil society plays an important role in the interpretation of collected data in providing context, and in anticipating data collection with the appropriate design of experimentation protocols. The cases reveal that civil society is very vocal in debates about the diffusion of innovation, but silent on other aspects of the innovation space. Cases investigated in this chapter do not make it possible to comment about the analysis of consensus spaces. Other data-collection protocols would be required to develop a research project on this specific aspect and go deeper into the analysis.

10.5 Conclusion

This chapter has focused on differences between contributions made by open labs to the Triple versus Quadruple Helix models for the governance of innovation. Cases considered in this chapter all relate in different ways to the interplay between knowledge, consensus, and innovation spaces to build the dynamics of interaction between the different Helixes and establish their role as boundary spaces and catalysts of innovation. This is our first contribution. The second point deals with the articulation between open labs and the dynamics of the Triple/Quadruple Helixes. When interactions between the Helixes do not exist before the installation of the open lab, gaining momentum will require the installation of trust and legitimation mechanisms meanwhile the execution of innovation projects develops. The ambition of such projects shall be adapted.

The third contribution shows the importance of appraising open labs in the context of the knowledge-based view (KBT) of organisations, to best appraise the links between knowledge, consensus, and innovation spaces in the dynamics of interactions. The reference to social learning cycles among stakeholders of the respective Helixes shows that managerial teams and experts of creativity or innovation management methods play a major role for the elaboration of effective projects. Section 10.4 suggests that it might be worth investigating further the boundary conditions explaining the elaboration of innovation projects as forms of boundary objects. The KBT approach also leads to the identification of specific modes adapted to interactions with civil society in the Quadruple Helix, with contributions with new insights in the creative process, actor of experimentation in innovation projects, and as co-creators of "boundary spaces". This chapter has identified the role played by civil society in the interpretation of data and in the preparation of the innovation projects, even though it remains difficult to provide any generalisation for their contributions to each subcomponent of the "boundary spaces".

Last, but not least, this chapter shows that the elaboration of innovation projects in the Triple and Quadruple Helix models of innovation benefit from the presence of open labs, because their teams enable innovation projects with the creation of knowledge, consensus, and innovation spaces. Many actors may play important roles in the different Helixes to orchestrate the respective contributions, but the cases investigated in this chapter show that teams present in open labs have always created the conditions for joint effectiveness. There is still much to investigate about complementarities between these spaces, but the chapter has already identified the prominent importance of the open labs' teams in the elaboration of knowledge spaces.

Bibliography

Bjørkquist, C., Ramsdal, H. and Ramsdal, K. (2015). User participation and stakeholder involvement in health care innovation – does it matter? *European Journal of Innovation Management*, 18(1): 2–18.

Boisot, M.H. (1998) *Knowledge assets: Securing competitive advantage in the information economy* (revised edition, paperback 1999). Oxford: Oxford University Press.

Carayannis, E. G. and Campbell, D. F. J. (2009). 'Mode 3' and 'Quadruple Helix': toward a 21st century fractal innovation ecosystem. *International Journal of Technology Management*, 46(3/4): 201–234.

Carayannis, E.G., David, F. and Campbell, J. (2006) "Mode 3": Meaning and implications from a knowledge systems perspective. In: E.G. Carayannis and D.F.J. Campbell (eds.), *Knowledge creation, diffusion, and use in innovation networks and knowledge clusters: A comparative systems approach across the United States, Europe and Asia.* Westport, CT: Praeger, pp. 1–25.

Carayannis, E.G. and Rakhmatullin, R. (2014) The quadruple/quintuple innovation helixes and smart specialisation strategies for sustainable and inclusive growth in Europe and beyond. *Journal of the Knowledge Economy*, *5*(2): 212–239.

Champenois, C. and Etzkowitz, H. (2018) From boundary line to boundary space: The creation of hybrid organizations as a triple helix micro-foundation. *Technovation*, *76–77*: 28–39.

Chesbrough, H. (2003). *Open Innovation: The new imperative for creating and profiting from technology*. Boston, MA: Harvard Business School Press.

Dzisah, J. and Etzkowitz, H. (2008) Triple helix circulation: The heart of innovation and development. *International Journal of Technology Management and Sustainable Development*, 7(2): 101–115.

Etzkowitz, H. (2003) Innovation in innovation: The triple helix of university-industry-government relations. *Social Science Information sur les Sciences Sociales*, *42*(3): 293–337.

Etzkowitz, H. and Leydesdorff, L. (eds.). (1997) *Universities in the global economy: A triple helix of university-industry-government relations*. London: Cassell Academic.

Etzkowitz, H. and Leydesdorff, L. (2000) The dynamics of innovation: From national systems and "mode 2" to a triple helix of university-industry-government relations. *Research Policy*, *29*(2): 109–123.

Etzkowitz, H. and Zhou, C. (2017) *The triple helix: University-industry-government and entrepreneurship*. London: Routledge.

Gibbons, M. (2000) Mode 2 society and the emergence of context-sensitive science. *Science and Public Policy*, *27*(3): 159–163.

Hasche, N., Höglund, L. and Linton, G. (2020). Quadruple helix as a network of relationships: creating value within a Swedish regional innovation system. *Journal of Small Business and Entrepreneurship*, 32(6): 523–544.

Heraud, J. A. (2017). Science and innovation. Chapter 4. In: H. Bathelt, P. Cohendet, S. Henn and L. Simon (Eds.), *The Elgar companion to innovation and knowledge creation*. London: Edward Elgar, pp. 56–74.

Höglund, L. and Linton, G. (2018) Smart specialization in regional innovation systems: A quadruple helix perspective. *R and D Management*, *48*(1): 60–72.

Ihrig, M. and MacMillan, I. (2013) The strategic management of knowledge, chapter 7. In: J. Child and M. Ihrig (eds.), *Knowledge, organization and management, Building on the work of Max Boisot*. Oxford: Oxford University Press, pp. 129–139.

Lettl, C., Herstatt, C. and Gemuenden, H. G. (2006). Users' contributions to radical innovation: Evidence from four cases in the field of medical equipment technology. *R&D Management*, 36(3): 251–272.

Leydersforff, L. and Deakin, M. (2011) The triple helix model of smart cities: A neo-evolutionary perspective. *Journal of Urban Technology, 18*(2): 53–63.

Lundberg, H. (2013) Triple helix in practice: The key role of boundary spanners. *European Journal of Innovation Management, 16*(2): 211–226.

Merindol, V., Le Chaffotec, A. and Versailles, D. W. (2022). The role of organization intermediaries in science-/techno-push versus user-centric approaches in health care innovation. *European Journal of Innovation Management*, forthcomming. https://doi.org/10.1108/EJIM-02-2021-0119 – online first Nov. 5th, 2021.

Paskaleva, K., Evans, J. and Watson, K. (2021) Co-producing smart cities: A quadruple helix approach to assessment. *European Urban and Regional Studies, 28*(4), 395–412.

Petersen, A.M., Rotolo, D. and Leydesdorff, L. (2016) A triple helix model of medical innovation: Supply, demand, and technological capabilities in terms of medical subject headings. *Research Policy, 45*(3): 666–681.

Ranga, M. and Etzkowitz, H. (2013) Triple helix systems: An analytical framework for innovation policy and practice in the knowledge society. *Industry and Higher Education, 27*(4): 237–262.

Sarpong, D., AbdRazak, A., Alexander, E. and Meissner, D. (2017) Organizing practices of university, industry and government that facilitate (or impede) the transition to a hybrid triple helix model of innovation. *Technological Forecasting and Social Change, 123*: 142–152.

Schiavone, F. (2020) *User innovation in healthcare: How patients and caregivers react creatively to illness*. Heidelberg, Germany: Springer Briefs in Healthcare Management and Economics, Springer.

von Hippel, E. (2005). *Democratizing innovation*. Cambridge, MA and London: The MIT Press.

Wachelder, J. (2003) Democratizing science: Various routes and visions of Dutch science shops. *Science, Technology, and Human Values, 28*(2): 244–273.

Afterword

Open labs, open innovation, and creativity: utopias and dreams to build meaning for the future

Michel Ida
Founder and former VP of Open labs, CEA

Translation by David W. Versailles

Is it still possible to move from "human-centric innovation" and "techno-push innovation" to "sustainability-centric innovation" preserving our planet and the climate? How is it possible to incentivise investments in science, research, and technological innovation, and the associated entrepreneurial behaviours? How to create desire and dreams about science and technology? How to make sense of technological progress in our modern civil societies? How to build a positive vision about innovation? How to generate greater commitment by research teams around progress while developing environmentally friendly innovation?

When voices advocate for a degrowth societal model and fear some sort of collapse, when the Covid pandemic and social movements shake up our geopolitical ecosystems all around the world, discussing technological foresight and the potential benefits of open innovation on society might represent some sort of provocation.

This Afterword promotes the point of view that science, technologies, and technological progress will make it possible to address the current exosystemic and societal challenges and propose solutions to current and future issues at stake. Progress, science, and technologies are especially important on a planet that is both facing increasing demographic challenges under the constraint of finite natural resources and confronting major climatic and environmental changes.

The beginning of the 21st century has revealed that the positive perception of science, and the acceptance of the benefits of technological progress, are no longer widespread in the population. Scholars and practitioners involved in innovation therefore think it mandatory to convince people about their positive impacts on our lives and our future. Short-term techno-push rationales for innovation management must be complemented with large and ambitious open innovation projects that will generate meaning, dreams, and commitment. Collaborative projects involving end users are required to anticipate behaviours, uses, needs, and practices that will prevail on medium- and long-term time frames. The innovation journey cannot take the people's acceptance for granted anymore; it shall dedicate specific efforts to build acceptance and create desire, emotions, and meaning.

Until the end of the 20th century, innovation was mainly managed in reference to the linear and techno-push model (Balconi et al., 2010). This managerial framework notably elaborates on incremental innovation stages, explicit gates, and risk minimisation. It is ruled by scientific and academic rigour. The 1980s saw two major evolutions: non-linear and user-centric innovation management models, and open innovation. User-centric and non-linear models introduce a "bottom-up" perspective that makes it easier to focus on end users' expectations and needs. This approach primarily refers to end users' perspectives to orient innovation processes. Over the last 40 years, open innovation managerial processes also developed in parallel to the science-push model and bottom-up processes, first in the USA during the 1980s and then in Europe. Henry W. Chesbrough's activities and publications about open innovation (2003) represent major milestones in this evolution, starting with his first activities at Berkeley. Open innovation is characterised by the now famous open innovation "*funnel*" (Chesbrough, 2003, e.g.: 44, 183, 189), and a pragmatic examination of interactions between in-house and external competency(ies). These aspects are both consistent with the science-/techno-push and user-centric approaches. Open innovation is now adapted and adopted by most companies. Non-linear innovation projects are characterised by a fuzzy and non-deterministic logic that was first rooted in design and the arts. This reference carries with it a great potential for disruption but it is also more complex to operate.

Even if the management of innovation remains totally different in the science- or techno-push versus non-linear/open frameworks, companies shall work on the meaning of innovation and its acceptability. Dreams and utopias represent mandatory references to generate commitment to innovative projects and to build agreement. This is important to support entrepreneurial journeys, attract investments, and motivate teams. Elon Musk or Jeff Bezos perfectly illustrate such dreams when they launch projects to move the boundaries of the possible in the space industry. Generating new interest for innovation requires the motivation of managers, investors, scientists, or entrepreneurs to deal with new frontiers that will in turn generate enthusiasm in the population. They need to believe in innovation projects and commit to societal change. This new way to consider innovation reshapes the links to investments and corporate strategies, as it has already changed the relationship to science and society. Innovation is no longer only based on rationality and the administration of scientific proof. It depends also on the installation of conditions for a strong twin commitment by companies (and their employees) and end users (and therefore civil society) for dream-bearing projects that also have an ethical content: protection of the planet, protection of the climate, inclusion of the most disadvantaged, reduction of consumption, less harm to humans.

Ideas Laboratory®, Atelier Art Science, Alps Design Lab, Human Lab, and the network of Open Labs installed by CEA Directorate for technological research after 2001, have all contributed to installing another perspective on the world and on innovation. These labs have been the pioneers of the "open lab" movement in France (Mérindol et al., 2016) that has been concretising the open and user-centric approach of innovation over the last 20 years.

Where do we stand today? What are the most emblematic instances of disruptive, open, mutualised, and "shifted" approaches that associate creativity, creation, design, architecture, anticipation on future uses and needs, scientific research, engineering, social sciences, political science, technological development, marketing, entrepreneurship, and citizens? What did they produce? This Afterword outlines the importance of utopias and dreams to encourage innovation all around the word. It describes how utopias and dreams framed the development of open labs towards innovation and creativity, first in the USA and then in Asia. The next section describes the vivid legacy of Bauhaus in pioneering initiatives led by CEA. The last sections identify new challenges and open perspectives for the future.

Utopias and dreams in the USA: Stanford's d.school and MIT's Medialab

Stanford University, in Silicon Valley, created an education programme in design as early as the 1950s to get out of traditional project management and retarget education on human-centric perspectives. By definition, design and architecture studies are both centred on uses and on actual behaviours. "Design thinking" emerged as a creativity method in the very same years to support this evolution, and transition from linear innovation processes controlled with gates to user- and human-centric innovation. The concept was made popular by David M. Kelley, CEO and guru of the IDEO company. Thanks to design thinking, IDEO became a leading company for creativity in 1991. To follow up and support IDEO projects, Kelley founded the d.school in 2004, Stanford design school, now a world reference in industrial design. Product designer Jules Sherman explains in the French newspaper *Le Monde* on March 25, 2016: "Everything was initially centred on fabrication processes. Now, we have human-centric designs". The d. school (later renamed the Hasso Plattner Institute of Design) is now funded by Stanford University and by industry.

At the d.school, education programmes are organised with one- or two-week-long seminars, and with semestrial residences. All programmes connect students with their faculties and industry. They instil strong incentives to collect data in real life, to observe and investigate new and original fields for research and innovation, to create prototypes, and to identify what influences behaviours. They repeat the design process until a solution satisfactory for end users is identified. Numerous successes can be listed for David M. Kelley and IDEO: the first laptop with a folding screen, Apple's first mouse, Palm V personal data assistant, etc. Today, IDEO works with cosmopolitan teams: scientists from human and social sciences, engineers, designers, biologists, cognitive psychologists, IT specialists, etc. IDEO does not limit itself to the design of products anymore.

Kelley very soon understood the impact of media and communication on his activities. In 1999, IDEO was highlighted in the TV show *Nightline* by the American broadcaster ABC, in a segment called "The Deep Dive: one company's secret weapon for innovation". Journalists were reporting about the mindset and posture manifested by IDEO experts who reinvented a supermarket shopping cart in five days. This famous 1999 video deeply reshaped the way to think about the management of innovation for innovators and companies. The most important decisionmakers and CEOs made this TV report their reference. Avant-garde designers and innovation directors are now imbued with this approach. Some of Kelley's famous quotes still have a major influence on designers. The first two sentences are taken from the *Nightline* TV report; the third sentence was written in *Creative Confidence*, the 2013 book by David M. Kelley and his brother Tom, now IDEO Managing Director.

- "Enlightened trials and errors succeed over the planning of the lone genius".
- "It's not organized [chaos]. It's focused chaos".
- "Failure sucks, but instructs".

This system fosters the emergence of teammates who commit to disruption while preserving their commitment to the company and its projects, who do not blindly follow rules and instructions, and demonstrate that they are at ease with flat hierarchies, individual autonomy, and collaboration.

In the USA, on the East Coast, the working atmosphere is different from that prevailing in Silicon Valley. In Boston, Nicholas Negroponte represents the emblematic figure for the management of innovation. With a master's degree in architecture earned from MIT, Negroponte created MIT Medialab at the crossroads between architecture, design, engineering, human, and social sciences. The focus on actual uses is immediately installed at the kernel of the activities with multidisciplinary creativity methods that were initially applied to innovation for man-machine interfaces. Each innovation project must have at least one artist or designer in the team.

Negroponte has soon become the innovation and creativity guru for Medialab and MIT. He also very soon understood the importance of media in this type of organisation: in 1992 he created the monthly magazine *WIRED*, that has today become one of the world references for the press on innovation. *WIRED* and similar magazines (such as *Red Herring*) have contributed to the wide promotion of Medialab in professional and non-professional audiences, and progressively generated industry acceptance. It is also possible to list a long list of societal successes, such as the "One laptop per child" initiative that benefited from tens of millions of US dollars granted by large American companies, and a long list of innovative products: the "less than US$100 computer" presented at Las Vegas CES in 2012, the MPEG4 norm, LEGO® Mindstorms,

IBM pointers, holograms for credit cards, Fisher-Price® Symphony Painter, etc. Today, Medialab has given itself the mission "to create a better future", which is similar to the strategic tagline prevailing in Ideas Laboratory® in the French CEA: "invent a desirable future together".

Utopias and dreams in Asia: from ITRI Creativity Center to HAX Accelerator

The transition towards new creativity and innovation management methods follows other rationales in Asia because of their original cultural background. The starting point of such an evolution can be identified about 15 years ago.

ITRI, the Taiwanese research institute in industrial technologies, is as a precursor of this movement for Asian countries. Wen-Jean Hsueh (VIATECH, May 25, 2005) created the ITRI Creativity Lab in 2004 and managed it until 2013 after spending several years at Caltech. As early as 2004, she was explaining the necessity of working with journalists, media, and artists to make people dream. She also wanted civil society to help pressure ITRI towards renewed innovation management pathways, and thus fight the prevailing engineers' mindset. She illustrates once again the importance of external communication and the perception of identities by the media.

Hsueh now devotes her time to interdisciplinary innovation, at the crossroads between human sciences, arts, and technology. She curated several international exhibitions on interactive technologies, covering technological innovation, design, marketing, and the creative arts. In 2006, she was a special guest at the Ars Electronica Forum in Linz (Arts Sciences) in Austria. She also became the first female Vice-President for Technology at the National Taiwan University of Arts in 2016, and the first female member of the board at the Chinese Institute of Engineers in 2012 (which was created more than a century ago). She is also at the forefront of research on 3D cameras and 3D displays.

Since these pioneering activities in Taiwan, other initiatives have been launched in Singapore, in China, in India, and in other countries in Asia. The research project commissioned by Bpifrance Le Lab to Paris School of Business newPIC chair (Mérindol and Versailles, 2019), and locally supported by the consulting firm *Innovation Is Everywhere* based in Singapore, shows that innovation platforms installed by large established companies prevail in Asia against "independent" open labs as analysed in this book. This research shows the rationales behind HAX Accelerator in Shenzhen in China, EVVX in India, Arkley Accelerator VC, and HUBA in Thailand, etc. Compared to France where the "non-profit" orientation is often present, "for profit" or "social business" rationales prevail for Asian open labs. In France, open labs mainly aim at developing new innovative capabilities and at handling change management. In Asia, open labs have an explicit articulation with large established companies and support their open innovation initiatives: they often source start-ups for these companies.

Both in France and in Asia, creativity methods (such as design thinking) and co-creation initiatives prevail. Open labs also devote specific efforts to the installation and facilitation of communities, with diverse profiles for contributors: entrepreneurs, employees coming from large companies, public institutions, non-governmental organisations, or non-profit organisations. Mérindol and Versailles (2019, 2017) show that two key drivers for successful activities apply to French and Asian open labs with the same level of importance: the dynamics of communities and the management of the physical space. It is important to manage conviviality and openness as catalysts of interactions in the open labs, and sources of proximity.

In these reports, Mérindol and Versailles reach the same conclusion about the importance of managers in open labs: the dynamics of interaction and collaboration do not emerge by some sort of miracle. Open lab managers must proactively design the space to foster conviviality and openness and facilitate individual and collective efforts. They must lead interactions "from the bottom", facilitate initial contacts, foster the organisation of events and "bottom-up" initiatives, and support their organisation. These ways of working remain original compared to the ones prevailing in large established companies. They are perfectly in line with exploration phases but require specific efforts to bridge with the exploitation phase in companies. This is the reason why specific efforts and initiatives are now in place in different companies (both in Asia and in France) around the management of open innovation in their ecosystems.

Utopias and dreams in Europe and in France: the legacy of Bauhaus and pioneering initiatives installed by CEA in Grenoble

The legacy of the multidisciplinary Bauhaus school (Shahabi, 1965) created in Germany in 1919 and closed by the National Socialists in 1933 might at least partly explain why the USA has initiated the different initiatives leading today to open, user-centric, and non-linear innovation management processes. After Hitler's accession to power in 1933, numerous artists and academics emigrated to the USA and there is no surprise that the main references behind the Bauhaus approach resurfaced in other forms 20 years later, in the main academic centres of the west and east coasts of the USA.

The Bauhaus school was active in architecture, applied arts, design, photography, creation of costumes, and dance. It is acknowledged as the precursor of design, following the Arts & Crafts Movement initiated by William Morris in the 19th century. Wassily Kandinsky, Paul Klee, Oskar Schlemmer, or Marcel Breuer manifested their adherence to the Bauhaus movement. It was at the origin of a genuine revolution in architecture and arts, with explicit concretisations in objects present in our everyday life since the 20th century, for instance furniture. Le Masson et al. (2016: 91) explain:

> *We analyse teaching at Bauhaus by focusing on the courses of Itten and Klee. We show that these courses aimed to increase students' creative design capabilities by providing the students with methods of building a knowledge base with two*

> *critical features: (1) a knowledge structure characterized by* nondeterminism *and* non-modularity *and (2) a design process helping students progressively "superimpose" languages on the object.* (Italics in original text.)

These authors explain the importance of generativity, a concept inherited from the Bauhaus movement: (strong or generic) generativity is a model associated with, or conditioned by, non-aligned knowledge structures. Le Masson et al. explain further (ibid., Section 2.3.1): "generic generativity can be obtained only with a knowledge structure without determinism and modularity". They established from Itten's and Klee's teaching that Bauhaus aimed to convey to students a capacity of creative design, a capacity to open "holes", to design "out of the box", and to "split knowledge".

This specific approach of open innovation resurfaced in the 1990s in Europe and in France, after the Second World War and the post-war period had totally occulted the Bauhaus philosophy for several decades.

In 1982, the French government decided to create ENSCI Les Ateliers, a design school that was intended to rejuvenate the Bauhaus philosophy in France. Post-modernism has also contributed to this new impulsion in a modified form, closely related to neorealism, raising the twin questions of ethics and of the uses of new technologies. These topics have been addressed by philosophers Markus Gabriel (2020), Bernard Stiegler, Dominique Christian (see Christian and Flamant, 2005; Christian and Maruani, 2020) and Thierry Menissier (2021). They all gave presentations to innovation communities and open labs. They have also been supporting innovation managers for several decades. The *White Book of Open Labs* (Mérindol et al., 2016) explains the importance of "shifted perspectives, creativity, and decentring" to see the world through renewed lenses. Indeed, this is a resumption of lessons taught by the Bauhaus teaching style about creativity, then adapted and developed at Stanford, Berkeley, and MIT. This is also, we are reminded here, the very posture enacted by an artist in the creative process.

French pioneers of the open lab movement are in Grenoble, at CEA and Grenoble Alps University.

Departing from traditional project management and adhering to the pioneer mindset described in the previous sections, CEA has organised its activities with a user-centric focus since the installation of its open labs in 2001: Ideas Laboratory®, Atelier Art Science, and Alps Design Lab. Ideas Laboratory® has also opened the path towards dense collaboration with the industry, generating more than €150 million in collaborative research projects between CEA and Renault, Bouygues, Colas, Rossignol, Essilor, ST Microelectronics, etc. Significant commercial successes were the outcomes of such research projects, for instance in the domain of movement detectors, graphical interfaces, video games, electric vehicles, etc. Ideas generated in Ideas Laboratory® also led to the creation of several start-ups: MOVEA, ISKNAMI, MORPHOSENSE, HYDRAO, etc. CEA's Ideas Laboratory® and Atelier Arts Sciences were also at the origin of conferences, exhibitions, shows, and events in the domains of innovation, sciences, technologies, arts and scenography, and non-standard professional fairs: "Ideas days", "Demo days", or EXPERIMENTA. Atelier Art Science initiated its activities in 2007 in partnership between CEA and Hexagone Scène Nationale Arts Science (the French Arts Sciences National Theatre). The French philosophers mentioned earlier have all been active in the open labs installed by CEA, either in the projects or in their boards. These initiatives are precursors for the Open Innovation Centre YSPOT inaugurated in 2022.

Mimicking activities organised by Stanford's d.school or MIT's Medialab, Grenoble hosted designers in residence in partnership with the design school ENSCI Les Ateliers Paris. This initiative lasted for ten years, between 2009 and 2019. It then transformed into the Alps Design Lab, and a team of designers directly integrated into CEA from 2020 onwards. Alps Design Lab serves as a federation of different academic French and international institutions, most notably in the domains of design, architecture, engineering, human, and social sciences. These institutions have sent hundreds of students for residences at CEA, or contributions to workshops: each year, 30 to 50 students contribute to projects defined by CEA and its industrial partners under modalities similar to the ones in place at the d.school in Stanford. Hundreds of concepts have been generated during these activities, and then exploited by the industrial partners or CEA Tech labs, for instance in the domains of augmented reality using smartphones, augmented reality glasses, mobility systems, the protection of public space, the management of emotions in future cars, etc.

Two major infrastructures recently emerged in Grenoble as part of this open lab network:

- The Maison de la Création et de l'Innovation, MACI (House of Creation and Innovation), inaugurated at the end of 2019 in Grenoble Alps University.
- The Open Innovation Centre YSPOT that is structured in two buildings. The first one, YSPOT Lab, inaugurated in January 2020, is intended to regroup teams supporting and facilitating innovation projects, to host a showroom and a fab lab for fast prototyping affiliated with CEA. The second building, YSPOT Partners, inaugurated in March 2022, hosts a village for start-up incubation operated by Crédit Agricole, and the already mentioned Atelier Art Science, Ideas Laboratory®, and the team managing and facilitating the CEA campus "GIANT".

Over two decades, large companies (PSA, Dassault Systèmes, Renault, EDF, Bouygues, etc.) or independent organisations (NUMA, Station F) have emulated the open labs born in CEA. Numerous open labs exist in Paris and its region (see Mérindol and Versailles, 2017) but they do not benefit from the backing of science, technology, and the potential for operationalisation that remain specificities of CEA's open labs. Continental France is always seeing increasing numbers of open labs. Lille Catholic University has the initiatives

Adicodes and *Humanicité*. *thecamp* in Aix-en-Provence is a typical example of an open lab where projects and contributors aim at dreaming about the future and work with shifted perspectives.

Managerial challenges

This book has deliberately not addressed specific questions about the decline of open labs in established companies, what the *White Book of Open Labs* (Mérindol, Bouquin, Versailles, et al., 2016) first coined as "bazaars" in reference to open-source software and innovation communities analysed by Eric Raymond in his essay *The Cathedral and the Bazaar* (Raymond, 2001).

Open labs are very similar to the "bazaars" coined by Raymond. Mérindol and Versailles (2017) have already pointed out that cathedrals and bazaars cannot follow the same organisational and managerial principles, that they associate with different mindsets and different profiles because they are respectively associated with exploitation and cost optimisation, versus exploration and disruption. Cathedrals must follow traditional project management and optimise the vertical management model. Bazaars adhere to horizontal models, aim at "splitting knowledge", expect to generate the unpredictable, and foster subversive behaviours that look for disruptions while preserving loyalty to projects and organisations.

The main challenge deals with the management of appropriate distance between the "cathedral" (the large company) and the "bazaar" (the open lab or the innovation platform). These elements have been described and analysed in the different publications by faculties working in the newPIC chair and were initially discussed in the *White Book of Open Labs* (Mérindol, Bouquin, Versailles, et al., 2016). This specific managerial issue does not only relate to the accommodation of outcomes of the exploration phase into the exploitation processes run by companies; it also points out that managerial processes sharply differ in the "cathedral" and in the "bazaar". This book has only focused on managerial processes applicable to open labs, as instances of "bazaars", with their specific challenges. Other considerations are specific to innovation platforms installed "in-house" by large companies.

Beyond the difference between "bazaars" and "cathedrals", it is important to draw a sharp difference between open labs committed to disruption and foresight, and laboratories that only work on incremental innovation, based on the extension of existing technologies or research findings. To best prepare for the future, to operate the transition from utopias and dreams to reality, open labs are required to handle disruptions with civil society. These aspects are already well acknowledged in North America, under the labels "prospective thinking", "future studies", and "forecasting". Management science scholars should think about these aspects and investigate the subsequent organisational and managerial issues. More than two decades of return on experience at CEA have made it possible to learn that "cathedrals", "bazaars", labs for incremental research, and open labs for disruptive innovation are not miscible in the same managerial methods. Mixing them into the same organisational pattern will most probably destroy their respective contributions. To state it in another way, one of the missions will prevail over the others, and the mindsets or behaviours will blend into a dominant pattern that will automatically frame the other activities.

Perspectives for the future

To close this Afterword, and this book, it seems important to highlight managerial challenges associated with "bazaars" as instances of "open labs". The different chapters in this book all stress in some way that the management of "open labs" shall foster the legacy of Bauhaus. This is the key for innovation! Here lies the strategic uniqueness of open labs as compared to "cathedrals" and to other organisations focusing on incremental innovation. To preserve open labs as the nest for the emergence of the most astonishing innovation and to build disruptive projects, managers of open labs, or "bazaars", should foster activities that "split knowledge", adhere to horizontal ways of working, and develop open innovation initiatives. With their focus on user-centric and sustainability-centric issues, open labs contribute to the building of meaning. They should be the origin of a new wave of engagement with innovation.

However, the future of innovation and the management of innovation makes it necessary to go beyond managerial issues and draw perspectives about the evolution of public opinions and the commitment of so-called "civil society" to innovation and technological progress.

After 20 years of anthropocentric and user-centric innovation, a major challenge has emerged in France and in other European countries. Even if this description somehow represents a simplification, the inflexion appeared some years ago after three main phases in the management of innovation: "science-" and "techno-push" innovation prevailed until the 1980s; "market-pull" innovation complemented it from the 1990s onwards; and "user-centric" innovation finally gained momentum during the two last decades. A new phase emerges now with the need for "sustainability-centric" innovation because the other pathways towards innovation cannot generate enough adhesion around the notion of "progress". The mobilisation of citizens in innovation projects represents a major motivation of the French public national policies "Territoires d'innovation" (innovation territories) initiated in 2017 but this is not enough. Four years after the initiation of this national programme, innovation projects must adapt and manifest a more global and more systemic vision of the world, aiming at generating greater sustainability and greater protection against climate change.

The acceptability of innovation has become a major issue in society. It turns into a threat to the diffusion of innovation and to the dissemination of technological progress. Illustrations for such demonstrations are easy to list in Europe with actions against the installation of 4G and 5G GSM antennas or against connected devices such as Linky electricity meters, or resistance movements against RNA vaccines during the Covid pandemic. Demonstrations develop on a global scale and multiply in all OECD countries. It is easy to guess that adoption issues will become even more challenging when dealing with artificial intelligence in the near future, or with quantum computing and quantum cryptography, or with new sources of energy such as hydrogen and the associated production and storage modalities. In parallel, ideas circulating on social networks always have stronger and more immediate impacts on public options. Whether these ideas are based on reason and rationality, or on political announcements, beliefs, and fake news, such ideas circulate the world and contribute to build public opinion with an equal status. The resulting symbolic halo (Simondon, 1960) surrounding technological innovation mixes elements of rationality with beliefs, relativism, and political short-term demagogy.

Lots of questions must be answered about innovation, that demand innovators to adopt a more comprehensive approach. Are we currently coming back to some sort of "magical thinking" (Levi-Strauss, 1966) as faced in other times? The increasing importance of fake news and beliefs against actual reasoning and scientific demonstrations represents a new issue to be considered in the management of innovation. Scientists and experts have now a decreasing influence about innovation. The value of technological progress is questioned before any rational debate can develop. Should scientists, experts, and specialists of technology remain in an external position of supplier of data to feed rational debates left to others, or take part in debates and contribute to shape public opinions about the diffusion of innovation?

There is no simple answer to these questions.

To reinstall a positive vision of expertise, scientists, engineers, and technologists should now work together with other human and social science specialists to fight partial perspectives and bring to the surface the ideologies underlying fake news. They should also team with specialists of the analysis of beliefs and with theologians, media experts, and analysts of influence in social networks, to name only a few, to promote rational debates considering both what is easy to see with superficial lenses and, also, what lies "behind the mirror". Ready-made responses hardly consider the global complexity of the issues at hand, for instance with the total costs of ownership or the long list of unintended and indirect consequences of phenomena.

The importance of open labs resurfaces here. As with all other organisations working on innovation, open labs now have to consider this societal inflexion. They can contribute to bringing solutions to these threats to our future. The open labs' openness to civil society, to citizens and to other forms of citizen representation (conventions, delegations, focus groups, etc.) makes them appropriate places to discuss the impact of innovation as a positive transformation of our everyday life, and experience the different aspects of a "sustainability-centric" innovation. Science and technology can bring progress in our modern societies, but it is of foremost importance for citizens and stakeholders to be parts of the different processes building a positive vision of innovation and technological progress. In open labs, citizens and users can work together with scientists, technologists, and innovators to transform their dreams and utopias into concrete solutions. They can understand the sustainability of proposed solutions. Open labs will become nests where innovation demonstrates its "sustainability-centric" contribution to the evolution of our world. We need to diffuse such demonstrations into public opinion and explain the complexity of phenomena. Open labs play a prominent role in building meaning for the future.

References

Atelier Art Science. https://www.atelier-arts-sciences.eu/English-47.

Balconi, M., Brusoni, S. and Orsenigo, L. (2010). "In defence of the linear model: An essay". *Research Policy*, 39(1), pp. 1–13.

Chesbrough, H. W. (2003). *Open innovation: The new imperative for creating and profiting from technology*. Boston, MA: Harvard Business School Press.

Christian, D. and Maruani, G. (2020). «L'érudition… dans la corbeille de l'ordinateur». *Hermès, La Revue*, 87(2), pp. 259–261.

Christian, D. and Flamant, S. (2005). «Narration stratégique : autour d'une intervention de récit assisté». *Revue française de gestion*, 159(6), pp. 283–302.

Experimenta. https://www.experimenta.fr/en/experimenta-the-expo/.

Gabriel, M. (2020). *The meaning of thought*. Cambridge: Polity Press, translated from the German, *Der Sinn des Denkens*, 2018, Berlin: Ullstein Buchverlag.

IDEO. https://www.ideo.com/about.

Kelley, T. and Kelley, D. M. (2013). *Creative confidence, unleashing the creative potential within us all*. New York: Currency Books, Crown Publishing Group (Penguin Random House).

Le Masson, P., Hatchuel, A. and Weil, B. (2016). "Design theory at Bauhaus: teaching "splitting" knowledge". *Research in Engineering Design*, 27, pp 91–115. https://doi.org/10.1007/s00163-015-0206-z.

Le Monde. (2016). March 25th. https://www.lemonde.fr/campus/article/2016/04/13/au-c-ur-de-lamythique-d-school-de-stanford_4900964_4401467.html.

Levi-Strauss, C. (1966). *The savage mind*. Chicago, IL: University of Chicago Press.

Menissier, T. (2021). *Innovations, Une enquête philosophique*. Paris: Hermann (see in particular chapter VIII, pp. 187–205).

Merindol, V. and Versailles, D. W. (2017). *Créer et innover aujourd'hui en Île-de-France : le rôle des plateformes d'innovation*. Projet de recherche financé par Bpifrance Le Hub, Innovation Factory et Paris&Co (avec des contributions d'Ignasi Capdevila, Alexandra Le Chaffotec, Nicolas Aubouin, Marion Desnost et Marianne Cohen pour la recherche de terrain). Paris: PSB.

Merindol, V. and Versailles, D. W. (2019). *Créer et innover aujourd'hui en France et en Asie : le rôle des plateformes d'innovation et des open labs d'entreprise*. Projet de recherche financé par Bpifrance Le Lab et Innovation Factory. Paris: PSB.

Merindol, V., Bouquin, N., Versailles, D. W., Aubouin, N., Capdevila, I., Le Chaffotec, A., Chiovetta, A. and Voisin, T. (2016). *Le Livre Blanc des Open Labs. Quelles pratiques ? Quels changements en France ?*, Travaux du groupe d'experts co-animé par ANRT / FutuRIS et la chaire newPIC de PSB. Paris: ANRT et PSB (Mars).

Raymond, E. S. (2001). *The Cathedral and the Bazaar: Musings on Linux and open source by an accidental revolutionary*. Beijing, Cambridge, MA and Farnham: O'Reilly Media.

Shahabi, A. R. (1965). *The Bauhaus and its contributions to design with suggestions for improvement of design in college industrial arts programs*, PhD Thesis, North Texas State Univ. Denton (TX), June, https://digital.library.unt.edu/ark:/67531/metadc163870/m2/1/high_res_d/n_03193.pdf.

Simondon, G. (1960). «L'effet de halo en matière technique: vers une stratégie de la publicité» In *Sur la Technique*. Paris: Presses Universitaires de France (2014), pp. 279–294.

Viatech. (2005). May 25th. https://www.viatech.com/en/2005/05/itri-creativity-lab-directoraddresses-vtf2005-on-innovation-and-technology/.

Annex A: List of open labs

Name	*Implantation (town for HQ, country)*	*General orientation and main topics covered*	*Creation*	*Website*
104 Factory	Paris, France	Incubator and accelerator for cultural and creative start-ups, inside the public institution, CentQuatre	2008	www.104factory.co m
La 27è Région	Paris, France	Lab for action-research programmes to transform new innovation policies	2008	www.la27eregion.fr
Active Ageing	Troyes, France	Living lab installed at UTT university, with a special focus on user-centric innovation applied to autonomy of elderly people	2013	www.activageing.fr
Allegro	Angers, France	Living lab installed at Angers university hospital, with a special focus on gerontology	2018	www.chu-angers.fr
Alps Design Lab	Grenoble, France	Open lab hosted by CEA Tech (Minatec) to work with design methods on use cases, prototypes, and demonstrators	2015	www.minatec.org
Arkley Accelerator VC	Warsaw, Poland	Incubator, accelerator, and venture capital supporting early-stage start-ups in the domain of deep techs grow from prototype to IPO, as part of PFR Ventures (supported by the EU)	2013	www.arkleybrinc.vc/ www.arkley.ventures
Artlab Digital Arti	Paris, France	Fab lab dedicated to digital arts	2011	Website not active
Atelier Arts Sciences	Grenoble, France	Open lab dedicated to activities mixing arts and sciences, created as part of CEA partnership with the Scène Nationale Hexagone in Meylan	2007	www.atelier -artssciences.eu www.yspot.fr

Bel Air Camp	Lyon and Villeurbanne, France	Open lab hosting a business community, coworking space, a fab lab, and meeting rooms, now with two locations (Bel Air Business and Bel Air Industrie)	2016	www.belaircamp.org
CEREMH	Saclay (Paris), France	Living lab dedicated to the experimentation of technological solutions for nursing homes, hospitals, and clinics	2008	www.ceremh.org
China Bee+++	Shenzhen, China, and four other cities	Coworking and co-living space offering services for residents' everyday life (bakery, fitness, concierge, etc.)	2016	www.beeplus.cc
CIC IT Lille	Lille, France	Open lab at Lille University Hospital co-branded with INSERM and the French ministry for healthcare, focused on e-healthcare, biosensors, and innovation in healthcare	2008	www.cic-it-lille.com
Cité de l'Economie et des Métiers de Demain	Montpellier, France	Open lab installed by Occitanie regional government to explore the future of jobs and support businesses in their transformation, with a special focus on social and societal challenges, and associated experimentation	2019	www.citedeleco.laregion.fr
cLLAPS	Paris, France	ICM's Living Lab, with the English name "Care Lab", the Healthtech living lab. ICM is the Paris Brain and Spinal Cord Institute	2007 (ICM)	https://institutducerveau-icm.org/en
Communitech	Waterloo, Ontario, Canada	Open lab and innovation hub helping tech driven companies to start and grow	1997	www.communitech.ca
Cowocat Rural	Catalonia, Spain	Association of rural coworking spaces	2014	www.cowocatrural.cat
CREATIS	Paris, France (and now Brussels, Belgium)	Open lab and flex office in the domain of cultural and media industries, initially installed at Gaîté Lyrique (Paris)	2011	www.residencecreatis.com
CRI (Centre for Research and Interdisciplinarity)	Paris, France	Now called Learning Planet Institute. Open lab focused on the learning-society revolution, both a digital innovation hub and a graduate and post-graduate school	2006	www.learningplanetinstitute.org/en

(Continued)

Name	*Implantation (town for HQ, country)*	*General orientation and main topics covered*	*Creation*	*Website*
Digital Village	Paris, France	Today a network of seven "villages" in France focused on coworking for digital workers; Paris's village is the only one that offers co-living and the organisation of events	2011	www.digitalvillage.com
Electrolab	Nanterre, France (Paris immediate neighbourhood)	Hackerspace and community of "makers" with a strong focus on prototyping and machines, plus teaching on technological issues and the use of machines	2012	www.electrolab.fr
Euratechnologies	Lille, France	Largest start-up incubator and accelerator in Europe, initially on deep tech and fintechs, now with specialist campuses in Roubaix (retail tech), Willems (green tech and "agri tech"), and Saint Quentin (robotics)	2009	www.euratechnologies.com
REVVX	Bangalore, India	Hardware Accelerator (pre-industrialisation) specialised in IoT and hardware products; in relation with the Indian Centre of Excellence for IoT and AI	2013 (2015)	www.revvx.com www.coe-iot.com
Found8	Singapore	Coworking space and flex office, located in the heart of Singapore's community of innovation, and offering direct connections to innovation leaders (events, etc.)	2019	www.found8.com
Les Garages XYZ	Saclay (Paris), France	Industrial innovation acceleration for deep techs, part of the Paris Saclay Hardware Accelerator project (see below)	2015	www.lesgarages.xyz
Hacking Health	Network head located in Montréal, Quebec, Canada	Global movement to improve healthcare related innovation with local communities ("chapters") and the organisation/facilitation of hackathons; more than 40 chapters today	2012	hacking-health.org
Hacking Health (Montréal)	Montréal, Quebec, Canada	Catalyst for collaboration and open innovation in health care, with HH "cafés", "design jams", hackathons, and "ateliers"	2012	hacking-health.org/fr/montreal
Hacking Health (Strasbourg)	Strasbourg, France	Or Hacking Health Camp (HHC) Specific chapter of the HH network	2013	hackinghealth.camp

HAX Accelerator	Offices in Shenzhen, China, and San Francisco, CA, USA	Hard tech venture capital, at the intersection between physical and digital (through its parent company SOSV, headquartered in Princeton, NJ, USA)	2012 and 1995 (SOSV)	www.hax.co
Health Care Factory	Paris, France	Open lab for digital innovation in healthcare	2012	www.healthfactory.i o
HUBBA	Bangkok, Thailand	Initially a coworking space, now expanded with innovation programmes for companies and start-ups into a start-up ecosystem catalyst and a community facilitator	2012	www.hubbathailand .com
I-Care Cluster	Lyon, France	Open innovation lab for healthcare, hosting a lab for ideation and experimentation; also in charge of running Lyon chapter for the Hacking Health network	2018	www.i-carelab.org www.hhlyon.org
ICI Montreuil	Montreuil, Paris area, France	Solidarity and collaborative makerspace (manufacture) for craftsmen, architects, designers, etc.	2012	www.makeici.org
Ideas Laboratory	Grenoble, France	Open innovation platform to share vision and new product exploration to make ideas happen, supported by CEA Minatec	2001	www.yspot.fr www .minatec.org
IDEO (OpenIDEO)	San Francisco, CA, USA	Open innovation platform focused on human-centric and design thinking approaches in healthcare	2010	www.openideo.com
iTMT (Institut TransMedTech)	Montréal, Quebec, Canada	Ecosystem for open innovation focused on the development of med tech, living lab	2017	www.polymtl.ca/tra nsmedtech/
ITRI Creativity Center	Zhudong, Taiwan, RoC	Applied technology research institute, now with a focus on three application domains (smart living, quality health, and sustainable development) and on intelligentisation-enabling technologies	1973	www.itri.org.tw
La Fabrique	Paris, France	Fab lab dedicated to the fashion industry, merchandising, etc. with a direct link to executive education programmes	2013	www.lafabriqueecole.fr
La Fabrique de l'Hospitalité	Strasbourg, France	Open innovation collaboration platform (living lab) between Strasbourg University hospital and local stakeholders, initially associated with the maternity hospital	2017	www.lafabriquedelh ospitalite.org

(Continued)

Name	*Implantation (town for HQ, country)*	*General orientation and main topics covered*	*Creation*	*Website*
La Paillasse	Paris, France	Hackerspace promoting the values of science as a means for social and societal innovation, and promoting interdisciplinary research and open (inclusive) science	2009	www.lapaillasse.org
La Ruche	Paris, France and five other French cities	Network of nine coworking spaces supporting projects in the domains of social innovation, and projects with positive societal impact	2008	www.la-ruche.net
Lab Santé Île-de-France	Paris, France	Innovation accelerator in the domain of healthcare, promoted by regional public institutions in health care, and by specialised R&D clusters	2016	www.labsante-idf.fr
The Tank	Paris, France	Coworking space and flex office space, dedicated to media, "thinkers", and entrepreneurs at the crossroads between societal, digital, and environmental issues	2014	www.letank.fr
Liberté Living Lab	Paris, France	Civic techs, and public good-related innovation. Self-introduced as "tech-related innovation with deep societal impact"	2016	www.liberte.paris
Lusage Lab	Paris, France	Living lab installed in Broca university hospital, specialised in gerontology (elderly people with cognitive impairment)	2010	www.lusage.org
Make ICI	Headquarters in Paris, France	Network of solidarity and collaborative makerspaces	2012	www.makeici.org
Make It Marseille	Marseilles, France	Coworking space and fab lab (makerspace), with storage space and photo studio	2013	www.makeitmarseille.com
Makesense	Paris, France, plus network of six cities (Mexico City, Lima, Dakar, Abidjan, Beirut, Manilla)	Coworking space (Makesense), incubator, community facilitator, venture capitalist (Seed 1) and education network, with a special focus on social and societal impact, and on social entrepreneurship	2010	www.makesense.org

NUMA	Paris, France	Formerly an innovation space with a special focus on digital innovation acceleration and venture capital organisation, now with a focus on executive education focused on a new generation of leaders	2002	www.numa.paris www.numa.co
Paris Saclay Hardware Accelerator	Saclay (Paris), France	Fab lab and makerspace focused on incubation, acceleration, pre-industrialisation, and production	2018	www.psha.fr
Plages Digitales	Strasbourg, France	Coworking space and makerspace for start-ups and innovation communities (in association with Alsace Digitale)	2012	www.laplagedigitale.eu
Station F	Paris, France	Generic incubator and thematic accelerator. Self-introduced as an "academy of entrepreneurship"	2017	www.stationf.co
Streetlab	Paris, France	Private living lab located in University Hospital Quinze-Vingt specialised in ophthalmology	2011	www.streetlabvision.com
The Corner	Brest, France	Innovation platform, incubation programme and organiser of hackathons, events, sprints, and education programmes on entrepreneurship with a special focus on the regional (Brittany) entrepreneurial ecosystem	2016	www.thecorner.fr
thecamp	Aix-en-Provence, France	Innovation hub/ecosystem dedicated to "positive innovation" and societal transformation, with different programmes for start-ups, SMEs, and large established firms, hosting a makerspace, an incubator, and hospitality capabilities	(2013) 2017	www.thecamp.fr
TUBA	Lyon, France	Living lab dedicated to urban innovation, to the digitalisation of urban life, and to digital education programmes; also hosts a coworking space	2013	www.tuba-lyon.com
Usine IO	Paris, France	Makerspace offering hardware innovation programmes (prototyping and small-scale manufacturing)	2013	www.usine.io

Annex B: Project commissioned by Bpifrance to the newPIC chair

Research project:

Taxonomy of innovation platforms in Paris and its region

Project commissioned by:
Bpifrance Le Lab, Innovation Factory, and Paris&Co

Execution date: May 2016 to January 2017

Principal investigators and project managers:
Valerie Mérindol and David W. Versailles

Research team:
Alexandra Le Chaffotec, Nicolas Aubouin, Ignasi Capdevila, Marianne Cohen, Marion Desnost, Nathan Afoumado, Etienne Cimetierre

Downloads: http://innovidf.newpic.fr

Description:
The project investigates the rationale and dynamics of open labs in Paris and its region, Île-de-France. The project produces a taxonomy supporting the elaboration of innovation maps, with a special focus on the originalities and complementarities between contributing open labs. It also analyses medium and long term strategic drivers applicable to them, the rationales for the labs' geographical implantation, and the main aspects of their business models.

List of open labs investigated in this project:
Creatis, Usine IO, Electrolab, ICI Montreuil, La Paillasse, Le Cargo, La Ruche (Canal), Liberté Living Lab, Le Tank, NUMA, Makesense Space, Digital Village, 104 Factory, Welcome City Lab (Paris&Co), The Family, Robotlab

Field research activities:
45 semi-structured interviews (open labs managers and founders, start-ups and small companies hosted in the open labs, large established companies using the open labs services); direct (non-participant) observation.

Research project:

Innovation platforms in continental France's main regional hubs

Project commissioned by:
Bpifrance Le Lab and Innovation Factory

Execution date: June 2017 to March 2018

Principal investigators and project managers:
Valerie Mérindol and David W. Versailles

Research team:
Alexandra Le Chaffotec, Nicolas Aubouin, Ignasi Capdevila, Océane Duyck, Salim Moulmaaz

Downloads: http://innovfra.newpic.fr

Description:
The project investigates open labs in continental France and checks the generalisability of the findings published in January 2017 about innovation platforms located in Paris and Île-de-France. The project investigates missions and ways of working, and interaction with the regions. It covers interactions with local and regional governments, investigates complementarities with local public and private stakeholders, and focuses on modalities used by open labs to support digital transformation. The project elaborates on the taxonomy produced in 2017, and confirms several drivers: the role of communities, the importance of reciprocity, the catalyst effect of the physical space, and the evolution of business models when public subventions phase out. The report confirms the role of open labs as brokers of knowledge and brokers of contents.

List of open labs investigated in this project:
Euratechnologies, Station F, The Corner, Metropulse, Herrera, Make It Marseille, thecamp, Alsace Digitale, Plages Digitales, La Fabrique, Le Tuba, Bel Air Camp, You Factory, TVT Innovation, plus an update on some open labs listed in the previous research report.

Field research activities:
50+ semi-structured interviews (open labs managers and founders, start-ups and small companies hosted in the open labs, large established companies using the open labs services); direct (non-participant) observation.

Research project:

Innovation platforms: France-Asia comparison

Project commissioned by:
Bpifrance Le Lab and Innovation Factory

Execution date: June 2018 to June2019

Principal investigators and project managers:
Valerie Mérindol and David W. Versailles

Research project operated in partnership with:
Innovation Is Everywhere (Singapore), represented by Martin Pasquier and Rachel Tan

Downloads: http://innovasia.newpic.fr

Description:
The project compares our analysis of France-based open labs with open labs and corporate labs located in Asia (China, India, Singapore, Thailand, Myanmar, Bangladesh, Malaysia) and some of their counterparts located in the USA. The newPIC chair performed interviews in China (Shanghai, Hangzhou) and partnered with Innovation Is Everywhere to operate interviews and visits under its scientific supervision. The project performed a stress test of the taxonomy produced in previous projects. It also investigated the impact of the specificities of ecosystems (different countries in Asia versus France) to analyse the role of boundary conditions in the dynamics of interactions in communities, of interactions with local public policy makers and with large companies, and the rationales prevailing in business models.

List of open labs investigated in this project:
REVVX, Airbus BizLab, SAP Startup Studio, Catalyst (Société Générale), China Bee+++, HAX Accelerator, ICI Beijing B&R International Co-Incubator, Nest.Bio, Plug and Play, TechCode, VAST, West Co-Village, X-Node, Atelier (BNP Paribas) Microsoft Studio, Found8, DBS DAX, Open Vault Fintech Innovation Center, Foundry (Unilever), Hubba, Phandeeyar, Startup Dhaka, Maybank Innovation Center, Arkley Accelerator VC Global, Founder Institute, Techstars Ventures, SOSV, and compared with several French Labs:
Bel Air Camp, Digital Village, Electrolab, Euratechnologies, ICI Montreuil, Make ICI, La Fabrique, La Paillasse, La Ruche, Liberté Living Lab, Makesense Space, NUMA, Station F, The Corner, The Family, thecamp, USINE IO, Wilco, Paris Saclay Hardware Accelerator, Les Garages XYZ

Field research activities:
70 semi-structured interviews (open lab managers and founders, start-ups and small companies hosted in the open labs, large established companies using the open labs services); direct (non-participant) observation.

Research project:

Innovation intermediaries in the healthcare sector

Project commissioned by:
Genopole, in partnership with Bpifrance

Execution date: May 2016 to January 2017

Principal investigators and project managers:
Valerie Mérindol and David W. Versailles

Research team:
Alexandra Le Chaffotec,
Aubery Thomas, Alexandre Blanc, Raphael Biroteau

Downloads: http://innov-sante.newpic.fr

Description:
The project investigates open labs as innovation intermediaries, and their roles in the healthcare sector and in deep techs. In this project, we compare Genopole (a major European hub for biotechs and gene therapies) with other open labs active in the healthcare sector, with innovation clusters, with innovation platforms focused on innovation and acceleration, and with global open innovation networks. We investigate how innovation intermediaries generate the intermediation function, and how to use it in the healthcare sector. We cover several dimensions: missions enacted by the intermediaries, their impact on the dynamics of communities and how to manage them, the dynamics of ecosystems (including demographics, geography, and agglomeration effects), governance issues, rationales for business models, and the ability to accommodate technological disruptions impacting the healthcare sector with user-centric innovation approaches.

List of open labs investigated in this project:
BioValey France (formerly Alsace BioValey), I-Care Cluster, IAR Cluster, MedicAlps, NextMed, Campus
GIANT, Genopole, ICM (Institut du Cerveau et de la Moelle Epinière), Lab Santé Île-de-France, Odeas
Laboratory™ as part of CEA's Open Innovation Center, thecamp, La Paillasse, WILCO, ActiveAgeing, Allegro, CIC IT Lille, Quattrocento, and the Hacking Health Network (with a special focus on Montréal, Strasbourg, and Lyon chapters).

Field research activities:
50+ semi-structured interviews (open lab managers and founders, start-ups and small companies hosted in the open labs, large established companies using the open labs services); direct (non-participant) observation.

Research project:

Cowocat Rural: the association of rural coworking spaces in Catalonia, Spain

Project commissioned by:
"Consorci Intercomarcal d'Iniciatives Socio econòmiques" (local consortium of socioeconomic initiatives); "Consell Comarcal" (County Council).

Execution date: 2014–2021

Principal investigators and project managers:
Ignasi Capdevila

Research team:
Ignasi Capdevila

Downloads: https://www.researchgate.net/publication/345342470_Coworking_Rural_a_Catalunya

Description:
In Catalonia, after the strong expansion of urban coworking, mainly in Barcelona, the concept started developing in rural areas from 2012, with the creation an initial space in Riba-Roja d'Ebre and extending later to a network of 15 spaces, distributed throughout Catalonia and now integrated into the Cowocat Rural project. The research project had a dual purpose. On the one hand, to summarise and analyse the development of the Cowocat Rural project to date, and on the other hand, to propose a methodological guide that could be used for territories that wish to launch a coworking project in rural areas.

List of open labs investigated in this project:
Zona Liquida, Espai la Magrana, Coworking Alfarràs, Espai Kowo, Nectar Conectar, CEI Les Borges Blanques, Zona Liquida, Espai Coworking Navata.

Field research activities:
27 interviews including managers and coworkers of eight Catalan rural spaces, as well as representatives of the Catalan government, regional coordinators, and managers of Cowocat Rural.

Index

Note: Page numbers in **bold** denote tables, those in *italic* denote figures.